T0168888

WHITETAIL SECRETS
VOLUME SEVEN

WHITETAIL SECRETS VOLUME SEVEN

SCENTS FOR SUCCESS

BILL BYNUM

DERRYDALE PRESS

Lyon, Mississippi

WHITETAIL SECRETS

VOLUME SEVEN, SCENTS FOR SUCCESS

Published by the Derrydale Press, Inc. under the direction of:

Douglas C. Mauldin, President and Publisher

Craig Boddington, Series Editor

Sue Goss Griffin, Series Manager

Lynda Bell Taylor, Administrator

Rick Carrell, Illustrator

Kirby J. Kiskadden, Designer

Cover: color photo by Douglas C. Mauldin

Frontispiece: Bill Bynum with a nice whitetail.

Frontispiece photo by Bill Bynum

Inquiries should be addressed to the Derrydale Press, Inc.,
P.O. Box 411, Lyon, Mississippi 38645, Telephone 601-624-5514,
Fax 601-627-3131

ISBN 1-56416-157-9

2 4 6 8 9 7 5 3 1

Printed in the United States of America
on acid-free paper.

DEDICATION

I am dedicating this book to two people who have proven to me that anything is possible. To my beloved mother, Thelma, who showed me there is always hope and nothing is impossible. Without your love and support a dream would not now be a reality. We have done it, doll!

Sharon, the love of my life. Your love and inspiration have proven to me that any brass ring is within our grasp. It is this love and devotion which will seal our love forever.

TABLE OF CONTENTS

EDITOR'S FOREWORD

More than a dozen years have passed since, at a barbecue following a dove hunt in Tennessee, I was introduced to a Tennesseean hunter named Bill Bynum. Bynum was uncharacteristically shy and soft-spoken that evening, but his message was clear: He thought he had something to say about hunting, and he wanted to know how he could get folks to hear him.

Over the years I've been approached by would-be and hopeful writers enough that I have a "canned" spiel in the back of my mind; it isn't that I don't want to help; it's simply that nobody can help much in such a quest. It comes from within. I gave him the spiel—but with Bynum the conversation was different than most. In the space of 20 minutes he convinced me that he did have something to say. I've believed in Bill Bynum ever since and he's never let me down.

I've always been grateful for that chance meeting in the Tennessee backwoods, for Bill Bynum has become the kind of friend that comes along few times in a lifetime. I've learned a great deal from him—and I know you will, too.

Tall, gangling, with a shock of unruly black hair that, even though Bill and I are of an age, hasn't started to thin or gray like mine, Bill has always reminded me of David Crockett, another Tennesseean whose roots he shares. If Bill thinks it ought to be said, he says it. If he thinks it ought to be done, he does it. And if there are whitetail in the woods, Bill Bynum will find them.

In the last decade Bynum has risen to national prominence as a speaker, writer, and all-around whitetail authority. He deserves whatever acclaim he has received—he's worked hard to attain it. However, unlike so many of today's self-styled gurus, Bill Bynum is real. As he puts it, he's a hunter first and a writer second. Or, as I heard him once shock an audience by saying, "We need more hunters writing and fewer writers hunting." Bill Bynum is a hunter—and the doggondest whitetail hunter I've ever known. No matter how rough the weather nor how heavy the hunting pressure, Bill Bynum will figure a way to outwit the deer.

He grew up hunting his Tennessee hills, where tough conditions and heavy pressure are a way of life. When the time came he left those hills and came back a decorated war hero. He doesn't talk about that much. Late one night he showed me a small memorial in the town square in Dresden. As has been tradition in Tennessee for centuries now, Bill and all his buddies went off to war together—and Bill was one of very few who came back at all.

Bill Bynum has no journalism nor English degree—and as I agonized over some of his early manuscripts it was clear that, as he says in these pages, his mind wandered through the hills and woods during his school days. But like everything he sets his mind to, when Bynum sets out to do something he does it. He's become a darned competent writer. And that's good news because, unlike so many of us, my friend Bill Bynum has something to say. As you'll find, it's well-worth listening to.

When it comes to whitetail secrets, Bill Bynum has more up his sleeve than any man I know. It gives me special pleasure to present this volume as part of our "Whitetail Secrets" series. Thanks to an early background in trapping and a uniquely inquisitive mind, the wise use of scents and lures is one of Bynum's specialties. I know you'll learn as much as I did in the pages that follow.

Craig Boddington
Lakewood, Colorado
June, 1995

INTRODUCTION

As a youngster I hunted, fished and trapped on an almost daily basis. The countless hours I have spent roaming and exploring the lands of western Tennessee were perhaps the happiest of my life.

Today I realize the days of youth were not wasted trying to unravel many of Nature's mysteries, mysteries which have guided me to understand many things pertaining to hunting. Mysteries which have led me to form opinions not only about hunting, but life itself.

Today I know from the bottom of my heart it has been the lessons I have learned which have inspired me to strive to teach others who share in the brotherhood of hunting. I know many of my brothers have not been as fortunate as I have. This I have learned from communicating with thousands of hunters across this great nation. It is through these communications I have also learned another aspect of life.

It is my opinion that friendship is one of the greatest rewards of a lifetime. If not for the devoted friendship of Craig Boddington and John Wootters, these words and this book might not have been possible.

It was the teaching and dedication of these two men which inspired me to take to the keyboard to write the many things I have learned while in pursuit of the whitetail deer.

Deer hunting has been away of life for me in many ways that only a hunter can understand. This knowledge can only be achieved from actual field experience.

The experience I have gained, just like my beloved friends, has been in the field. Far too often hunters are misled in the difference between experience and realization due to today's "grab a $buck$" world.

In the pages that follow, I express my personal feelings and follow the command of my teachers. I can only promise each page will hold true to the words "Your readers come first!" and "Tell it like it is and damn the consequences!"

In summary, to all who read this book, I am a hunter first and a journalist second.

Good luck and good hunting!

Bill Bynum
Dresden, Tennessee
April, 1995

CHAPTER ONE

UNDERSTANDING SCENT

My first recollection of scent and scent usage came at an early age. I was perhaps nine or 10 years old when I began trapping various furbearers for some extra cash. The money collected from fox, raccoon, and mink pelts helped support my hunting habits. Without the money from the pelts the chances of a new shotgun or deer rifle were pretty slim.

With little instruction on the art of trapping, most of what I learned was from trial and error. Errors meant a varmint had neglected to enter the jaws of my trap—and without the varmint entering the trap I had no fur and no money. This resulted in many sleepless nights trying to figure out what had prevented me from catching the animal.

Like most things in life I learned many important lessons from my errors. The dozen or so steel traps were checked every evening after the farm chores were completed. With luck, some critter was skinned by the light of a Coleman lantern in the old barn that served as my fur shed and "laboratory." This laboratory was where I experimented and developed some very foul concoctions I proudly proclaimed as trapping lures.

Among the shelves of the barn an assortment of labeled jars could be found. The labels read such various words as "sweetflag," "mink glands," "raccoon urine," and other unpleasantries. It was from these jars I created various blends

The whitetail's nose is like a magnet, receiving virtually all the scent molecules in the air currents. (Photo by Judd Cooney)

which lured the animals to my awaiting traps. It was also through using these blends that I learned many other things pertaining to how important scent was in communicating with animals. That's right, I said "communicating" with the animals.

Years of experience have enabled the author to harvest nice whitetails like this one.

Many hunters fail to realize we are forming a type of communication network with scent. Scent, like vocal communication, is a form of information we send to the animals. If we transmit the proper scent message, a positive result can be expected. If an improper message is relayed, the obvious negative result will occur.

Negative results usually occur from inexperience, in my opinion. Hunters simply fail to realize what they are doing when they attempt to relay a scent message. Often this is committed from improper instructions on the way the message should be transmitted. This is due to various factors which will be discussed throughout this book.

In the pages which follow I will try to explain the importance of properly transmitting scent. I will also voice my opinion on various factors pertaining to scents and their usage. My opinions and beliefs are based on the my experience in dealing with scents over the past three decades—decades which have proven to me the benefits of using scents for attracting and repelling deer.

HOW SCENT WORKS

In the realm of modern-day deer hunting, I feel the subject of scent is perhaps the most complex. Some hunters believe the subject of scent has no bearing on hunting success; many other hunters believe scent will be a determining factor at the end of the hunting day. In short, the subject of how scent affects deer hunting is somewhat of a controversial issue.

This issue, in my opinion, is not a question at all but a hard cold fact which will determine our hunting success. Unlike some hunters I have learned to respect the nose of the whitetail deer.

The author's knowledge of scent usage began with trapping at an early age. Here the author's son follows in his father's footsteps.

The reason I respect the whitetail deer's nose is because it has foiled many hunts for me, after I had spent countless hours plotting what I thought were almost foolproof plans, only to go home empty-handed. I went home empty-handed because of a slight change in the air currents which carried the invisible microscopic things called molecules to the deer's nose.

The dictionary defines molecules as being the smallest particle of a substance, whether an element or a compound, which can exist independently and still retain its physical and chemical identity.

In plain words, molecules identify whatever may exist. Whether it be a tree or a hunter, the subject consists of molecules. It will be the amount of these molecules the deer's nose receives which will determine the amount of *scent* that will be detectable. Scent is defined by the dictionary as "odor, fragrance; or an odor by which a person or animal can be traced."

The whitetail has more than 10 scent-producing glands, so clearly scent plays an important part in all aspects of the whitetail's life. (Photo by Judd Cooney)

With these definitions the words "scent molecules" should become important in a hunter's vocabulary. I can assure you they are in mine, because it is these scent molecules which will determine how a deer may react to any given situation. But before the deer can react, it must receive the scent molecules; therefore, I feel it is important for hunters to understand the aspects of scent and how it is transmitted.

This is the reason I advise hunters to use their imagination when thinking of how scent really works.

Imagine the nose of the deer as a super magnet and scent molecules as tiny steel ball bearings. If you have ever played with a magnet you know how quickly the force of the magnet retrieves metal objects. This is the same way a deer's nose retrieves scent molecules. Then the deer can identify the source from which the molecules originated.

ALL SCENT IS NOT CREATED EQUAL

To me, scent is the most personal element of any existing thing. Scent is a form of identification which not only distinguishes the source, but its actions as well. An excellent example of this is the skunk.

A hunter can smell a skunk while it is quietly sleeping inside a hollow log. If the hunter decided to sneak up on the skunk and shout inside the log the scent would probably increase dramatically. This would be due to the animal's reaction at being frightened.

Due to the sudden fright, various scent-producing glands within the skunk would become activated and produce higher amounts of scent molecules.

The increase in the scent molecules would then be transmitted by the air currents, and thus, the number of molecules the hunter would receive would be many times greater. This increase is why wise hunters should avoid sleeping skunks.

I also think this example explains why the number of molecules determines how scent is identified.

DO WE REALLY KNOW ABOUT SCENT IDENTIFICATION?

Identifying various scents and trying to relate to their meaning is a very complex subject. Numerous biologists around the world try to unlock these mysteries every day. Often hunters hear about some of the research gathered from these biologists. Often as not, this information is related in a manner that promotes some sort of hunting product which is supposed to provide hunters with better hunting results. Oh, boy, here we go again. The ultimate deer lure has been developed!

There is no question biologists can relate many aspects of how various scents affect deer and their behavior. This is especially true of the animals which were used in the research project, as the animals were cared for in a controlled environment—an environment which is totally different from that of wild animals.

There is no doubt in my mind they can tell someone how deer number 6127 reacted to the smell of garlic in its food. The scent made deer number 6127 snarl and run away. This reaction possibly made deer number 6127 produce some form of scent. This is based on the reaction of deer number 3251.

When deer number 3251 came to the feeding area, it reacted strangely. Deer number 3251 raised its tail when it smelled where deer 6127 had stood two hours before and became excited.

Was the excitement of deer number 3251 due to the scent of deer number 6127 or the scent of the garlic? Is this

Whitetail communicate in many ways; scent is just one avenue of communication, but it's an extremely important one. (Photo by Judd Cooney)

a major breakthrough in deer communication? I hardly think so! I do believe scent is a very important element in the communication network of the whitetail deer. I also believe there are some biologists who have formed some sound opinions. The point I am trying to make is captive animals react differently! I formed this opinion after having dealt with both captive and wild animals for numerous years.

I simply believe we have a long way to go before this network of communication will be unraveled, if ever! I am not saying the information gathered from captive animals is not important, I just don't think it will help hunters very much.

Big bucks come few and far between. The seasoned hunter is always prepared for that one chance in a lifetime!

Hunter hunt wild deer which are controlled by their habitat. It will be the habitat which will influence the habits of the animals; therefore, the more a hunter understands the habitat and the habits of the animals, the better. I firmly believe hunters must first understand all of the elements of habitat (food, cover, other animals, etc.) before a hunting strategy can be administered. The reason for this is, the habitat must supply the deer with all its basic needs.

The basic needs of a whitetail are simply eating, sleeping, and reproducing the species. In many instances scent will be an important factor pertaining to a hunting strategy. How the scent transmission is received by the animal can determine the outcome of the hunt.

The reason for this is various scent transmissions can represent some of the deer's basic needs. In most cases, the scent of either food or sex will be the most common.

This is why it is important for hunters to understand what I have previously written: We must represent something which is natural to the animals to avoid alarming them. I feel most hunters alarm the deer without knowing it. This done by not respecting the intelligence level of the animal.

HOW DUMB ARE DEER?

Over the years I have heard many hunters refer to deer as being dumb animals. Usually these hunters are standing around voicing their opinions after a day afield. In most cases these same hunters have spent the day hunting without bagging a deer. So, my friends, if the deer were so dumb why didn't you harvest one?

If there is one thing I have learned about whitetail deer it is to never underestimate them! This is very true when you begin using scents against them. Respecting the olfactory capabilities of a deer is also a form of respecting the animal's intelligence.

Hunters who learn to respect all of the deer's senses will dine on a lot more venison than those who do not. I have witnessed some pretty clever things deer have done.

Things such as seeing something odd and swinging downwind to make positive identification. It has been instances such as this which have brought me to an important conclusion about this remarkable animal: No matter what a deer may hear or see, it will always believe its *nose*. This is why I feel it is important for hunters to be scent-conscious at all times.

Seasons hunters know the buck of a lifetime will come only one time. Hunters who are scent-conscious at *all* times will enjoy the moment for a lifetime!

CHAPTER TWO

SCENT AND THE DEER

The chill of the late October air was dissipating with the warming sunlight. Hours had passed since I had climbed into my treestand. From my treestand I could overlook a dried creek bed and hardwood forest. Behind me was a dense cedar thicket I knew was a bedding area for the deer. I felt I had an excellent location for ambushing a trophy buck. Sadly, while I had been in the stand, I had seen only one lone whitetail doe that morning.

The doe had walked within mere inches of my tree as she headed for her bedding grounds. At no point in time did I make any movement or sound. I simply did not want to take any chances of alarming the doe because I had seen a splendid 10-point buck only a few days earlier less than 10 yards from my stand. I was confident the buck would soon adorn my wall.

I had taken every precaution I could think of in the prevention of making my presence known to the buck. Unfortunately the only memento I have of the 10-pointer is a valuable lesson.The lesson began when I spotted the buck coming towards my stand. With each step the deer took, my pulse quickened and the excitement increased. Like a kid at Christmas, I felt my wishes were about to be answered. I became even more excited when the buck crossed the creek bed and begin his final approach towards me. Everything was fine until the buck suddenly stopped!

From my treestand I could see the buck had become very nervous about something. I could also see the deer's attention was focused towards my direction. My first thought concerning the deer's actions was the wind had suddenly shifted directions. This thought was quickly eliminated with a glance at my indicator. I was feeling uncertain when I noticed the buck sniffing a small tree. Why was this, I wondered?

Long moments passed as I watched the buck smelling the tree. I was becoming baffled by the deer's actions when I realized something. I had used the tree to climb the bank of the creek that morning.

Seconds slowly became minutes as the buck stood motionless looking in my direction. Then, as if the buck could sense me, it turned and trotted back in the direction it had come from. A direction the buck knew would be safe as it had just traveled it. With its departure, my heart dropped as my brain began to engage.

After several minutes of thinking about what had transpired, I realized I had placed my scent on the tree—a lesson which I have never forgotten. Sadly, many hunters fail to realize how trees play an important role in scent distribution.

Like the tree which transmitted my scent to the deer, trees are used to transmit scent to other deer, also. The most common of the tree transmitters are rub trees.

RUBS ARE COMMUNICATION TRANSMITTERS

Rub trees are trees which have been marked by deer. Rubs are created by the buck rubbing tree bark with its antlers.

This, in turn, will remove much or all of the tree bark, presenting a very visible attractant. After the rubbing the

The author believes hunters often alert their presence without realizing it. (Photo by Judd Cooney)

buck will place its personal scent from its forehead gland. This scent informs other deer of the buck's social standing within the area. In many ways a rub is like a personal business card for the buck. This is the reason rubs are impor-

A buck rub is a place of information for all the deer in the area. This trophy buck was taken while working his rub line.

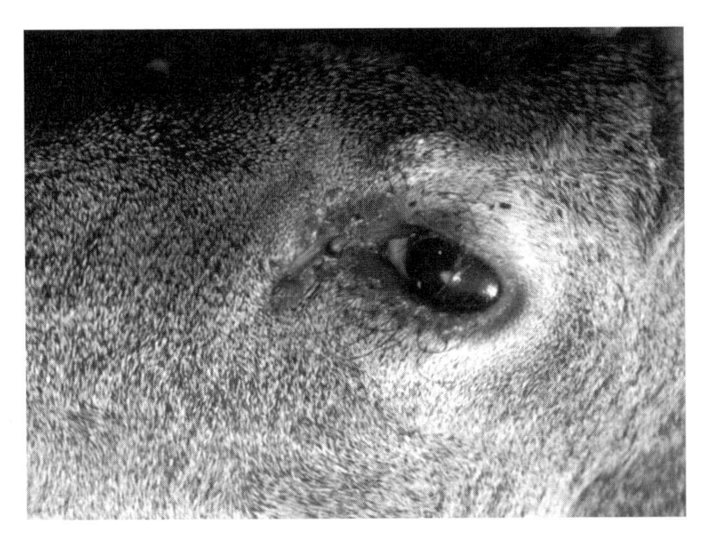

Much of the deer's personal scent is transmitted from the pre-orbital gland.

tant to the buck. These rubs will help in determining the pecking order for the upcoming breeding period.

Another scent which is often placed on the rub is the scent from the pre-orbital gland. The pre-orbital gland, often called the tear duct gland, is located below the eye of all deer. From the pre-orbital gland, small amounts of scent can trickle out and deposit the deer's scent when needed.

The pre-orbital gland is a gland which I feel is really not understood by man very well. I have read numerous words about the purpose of the pre-orbital gland. Some people feel the importance of the pre-orbital gland is strictly for doe/fawn contact. They believe the doe smears her pre-orbital scent upon the fawn while nuzzling. This is a very sound theory in my opinion as most fawns are identical in appearance. Therefore, the scent placed upon the fawn would be a form of identification to the doe.

Another theory of the pre-orbital gland takes us back to

the subject of rubs. The theory is that the pre-orbital gland is used to mark the animal's territory. This theory is really more than an educated guess, in my opinion. It's a fact! There have been numerous occasions where I have watched bucks create rubs and rub the tree with the side of their faces. This indicated to me the buck was using the pre-orbital gland to mark its rub.

Another factor which makes me believe the theory pertaining to territory marking is I have found small amounts of a waxy solution on rubs. This same type of substance is often present on the pre-orbital gland. This waxy substance has played an important role in one of my favorite hunting techniques, which we will review later.

TARSAL GLANDS: THE BUCK'S CALLING CARDS

The first dealing I ever had with tarsal glands came when I was about 11 years old. Like most young hunters I wanted to shoot the biggest buck in the country. My knowledge of whitetail deer was somewhat limited at the time. I had heard of a few men who had a reputation for achieving my dream on a consistent basis. It would be these men whom I would pester until their secrets for success were mine.

After much persuading, I finally talked one of the gentlemen to come and hunt on my farm. The gaining of new ground inspired the ol' boy to give me some pointers on trophy buck hunting. His knowledge of deer hunting was quickly proven when he hammered one of my big 10-pointers the first time out! This, I felt, was a just payment as he showed me his secret of hunting deer.

The secret was conveyed to me after we had dragged the large buck to my friend's truck. With the buck resting in the bed of the truck, my friend took his knife and began cutting the back legs of the deer. As he cut he informed me these were the tarsal glands of the buck. "The tarsal glands were the key to luring sly old bucks from their beds," he stated. "Simply hang these up in a tree and a buck will come a'running!" With these words of wisdom I proudly took the smelly glands and wrapped them in a plastic bread sack.

BYNUM GETS SERIOUS

Daylight found me sitting in the forks of a large oak tree the next day. My excitement had kept me awake most of the previous night with the thought of monster bucks, which I knew would come from far and near to smell my secret weapon. Hours slowly passed by, and the only thing resembling a deer hung from small tree. Like two pieces of dead meat (which they were), they just hung there. No monster bucks came crashing through the woods to smell my secret weapon. In fact, the only thing which showed any response to the hanging glands were a bunch of flies!

The remainder of the day was spent with me sitting in a tree staring at tarsal glands. It was not until sundown did I notice some movement out of the corner of eye. Approximately 50 yards to my right I saw what I knew was a dandy set of antlers. With this I slowly began shifting myself into shooting position.

Knowing I had to brace myself for the recoil of the twelve gauge slug, I wrapped my legs around the big limb. In moments I knew I was about to become a master deer hunter as the buck kept coming. The sight of the approaching deer played havoc on my nerves. I could only think about how I would become famous due to my newfound secret. Hunting big bucks was now about to become a snap as my first victim was about to fall.

Perhaps it was my thinking which interrupted the buck's pace, or maybe it just heard the pounding of my heart, I don't know. I do know the buck suddenly stopped and shot its nose straight up in the air. With this movement I knew it was now or never and I drew a bead on the buck.

Today the memory of that splendid eight-pointer still causes me to question the use of tarsal glands. Like many other deer I have shot while using extracted glands, I question whether they work or not.

There have been numerous times I have used tarsal glands and tarsal gland products. During these times I have seen deer react in a manner such that I feel they were responding to the scent of the glands. There have also been times when I felt the scent of tarsal gland repelled the animals.

Today, after years of using tarsal glands, I have developed some theories about tarsal glands. I base these theories on knowing more about the purpose of the gland and hunting experience.

WHAT IS PURPOSE OF THE TARSAL GLAND?

The tarsal gland is a true external gland located inside both hind legs of the deer. The exact placement of the gland is at the heel/foot junction of the leg. In many cases the tarsal gland will appear much darker in color than the surrounding leg hairs.

I have read where some biologists felt the tarsal gland is the deer's most important scent gland. They feel the tarsal gland produces the scent which relates the social dominance of the deer.

The tarsal gland also plays an important role in the reproductive activities of the animals, as well as a specific form of communication.

The author has never harvested a buck like this one using a tarsal gland.

This mode of communication is transmitted when urine from the deer flows onto the tarsal gland. It is with this mixture the various communication messages are released. With this important information we understand the tarsal gland produces a highly important scent for the deer. This knowledge has helped me in developing my own theories in the use of tarsal glands for hunting.

SHOULD WE OR SHOULD WE NOT?

There is no question in my mind that some readers will not agree with this. This is fine with me as I believe that the most important thing a hunter can take to the woods is *confidence*. Therefore, if you have confidence in tarsal glands, may the force be with you. Personally I feel the use of tarsal glands is not all that great. The reason I feel this

Today there are various products from the various glands of the deer. The author believes confidence is the key to harvesting deer

way is based on this: the tarsal gland is exactly what its name says it is, a gland! Like any gland it functions from the life of the creature it is attached to; therefore, when life no longer exists, the gland cannot function. Think about it!

THE INTERDIGITAL GLAND

The interdigital gland is located between the toes of each foot of the deer. Unlike some glands, the interdigital gland begins functioning at birth. This gland is used to place the personal scent of each animal on the ground. The interdigital gland is a highly fragrant gland and releases scent each time the animal takes a step. This is a very important gland as it aids in deer tracking each other. This allows female deer to track their young and males to track females during the rutting period. It is during the rutting period the hunter may become interested in the importance of interdigital gland.

GLANDS IN A BOTTLE

Once again, some commercial scent companies have come to the aid of us hunters. Today we can recreate deer walking through the woods with INTERDIGITAL GLAND LURE. These products have been created to give the hunter an edge in harvesting deer. In some cases I am sure they have been a factor in hunter success. In some cases they have probably have had an influence in whitetail deer success. This is the reason I have the same emotions towards interdigital gland lures as I have towards any gland lure.

HOW DEER GLANDS AFFECT THE HUNTER

Without question, few hunters give much thought to the various glands of deer. This is possibly due to the hunters caring more about shooting the deer than the biological make-up of the animal. Personally I, too, have felt this way until I began exploring the ways of the whitetail deer. Today I contribute much of my success to having knowledge and understanding pertaining to the various glands.

In this chapter I have reviewed what I felt were the most common glands by which hunters are affected. I have also voiced my opinion regarding various aspects of scent glands. I hope that I have made one thing very clear about all of the glands of a deer: they are part of the animal's way of life.

Wise hunters realize the more knowledge they have about the animal's life, the better they can hunt the animal. With over 10 external scent-producing glands on the deer's body, I would say scent is a major factor in the life of a deer!

CHAPTER THREE

SCENT AND THE HUNTER

Like most hunters I venture afield to enjoy myself and hopefully harvest a deer. Much of the pleasure I receive is from trying to unlock some of the many mysteries pertaining to the whitetail deer. For decades I have been fascinated by this creature and find the fascination growing each day. I have also gained a fascination for those who hunt the deer. These are creatures (deer hunters) who are more unpredictable than the animal they hunt.

Deer hunters are fascinating because most are creatures of habit and don't even realize it. They climb from their warm beds hours before daylight in the dead of winter. These same creatures then sit for hours in bone-chilling temperatures. It is also these same creatures who often wonder why they are rarely successful at harvesting a deer. They simply cannot understand why the deer avoid them while they are hunting. This is because they fail to realize the mistakes they make.

There is no question in my mind the human brain is more capable of reasoning than the brain of a deer. Unfortunately we hunters sometimes simply forget to engage our brains before going afield!

I have discovered this to be very true of a good hunting friend of mine. In many ways he simply defeats his purpose days before he goes hunting. How can one do such a thing? Simple, by the foods he eat!

DEER HUNTING AND YOUR DIET

The degree of seriousness is diverse among deer hunters. Some hunters try desperately to leave no stone upturned in planning their strategies. I often find myself doing things which seem ridiculous at the time I am doing them, only to discover their importance in the field. One such thing is my pre-hunt diet. Unlike my hunting buddy, I avoid foods containing high choloresterol levels. Foods which contain chloresterol produce body heat. This, in turn, increases the amount of human scent we produce.

The increase may not seem to be much to some hunters, but it is an increase. To a serious hunter, the increase could play an important role in our hunting plans. This is the reason I avoid these foods within a 24-hour period of hunting.

Instead of eating a big T-bone steak for dinner, I fill up on fish and salad. While my good friend is shoveling down bacon and eggs for breakfast, I have pancakes or Captain Crunch.

YOUR BREATH STINKS TO DEER

Something that has always amazed me about some hunters is their *breath*. Have you ever met someone and discovered their breath had the fragrance of diesel fuel? I have, and wondered how the deer interpreted the scent.

I don't believe a whitetail deer understands what diesel fuel is, but I believe they realize it is something not common to the woods!The scent from our breath can be reduced by practicing good dental hygiene. I have found brushing my teeth and rinsing with a mixture of baking

Hunters who are always striving to increase their odds are usually successful. (Photo by Judd Cooney)

soda and water reduces breath odor. This, in turn, not only makes my mouth feel fresher, but increases my confidence.

•

SPICE IS NOT ALWAYS NICE

Like many hunters I enjoy eating spicy foods. The taste of garlic, sage and other spices makes eating more enjoyable. Unfortunately, some of these spices can reduce our chances of eating venison. Spices such as garlic and other strong smelling spices can linger for many hours. I try to avoid these spices before going to the woods. I learned this one day while hunting with my good friend whom I previously mentioned.

The use of a scent eliminator can reduce odors within the hunter's hair.

We had been bow hunting in some unusually warm temperatures. Upon returning to the truck to head home, I noticed a strange odor. At first I could not distinguish the aroma lingering inside the truck. After a few minutes of playing of "What's that smell?" I asked my friend, "Have you been eating garlic?" His reply was, "How did you know that?!"

When I told him I could I smell it, he had a very puzzled look on his face. He informed me that he brushed his teeth and checked his breath afterwards. He said that he had not detected the scent in any way. I thoroughly believed him as I knew he was trying desperately to impress the lady he had been to dinner with night before.

After listening to my friend relate what he had eaten I discovered he had consumed a high amount of garlic. I informed him that various strong smelling spices could be transferred through our bodies. The scent of these spices could then later be released through the pores of our skin.

I also informed my friend that if I could detect the scent,

The author believes many hunters fail to respect the nose of the white-tail deer by eating spicy foods before hunting. (Photo by Judd Cooney)

I figured a deer could, too. Like many hunters, he paid little attention to my philosophy. He told me that I took the factors of scent too seriously and should relax and enjoy some of life's little pleasures.

The following day I took my friend's advice and enjoyed

some of life's pleasures. The pleasure was dragging a splendid six-point to the truck. I also enjoyed the pleasure of informing my friend I had now filled my bag limit. Now I could go to the market and get some fresh garlic to prepare one of my favorite venison recipes. In short, there is a time and a place for everything!

HAIR CAN SCARE

It is often funny how we sometimes discover factors which influence our thinking. A few years I encountered a skunk which had wandered into one of my coyote traps.

As would be expected, I carried the fragrance of skunk home with me. I spent a very long time in the shower trying to remove the scent from me. I felt I had done a marvelous job and was ready for my dinner date.

That evening when I met my date I noticed she was somewhat reluctant to sit beside me. After some time I asked if there was something bothering her. It took only a few seconds for her to inform me I smelled like a skunk. This was somewhat shocking as I had used scented soap and a heavy amount of cologne. Needless to say, I explained what had happened and apologized.

The next day was spent checking my traps and preparing for the upcoming evening. My enthusiasm was quickly deflated when my date informed me that she could still smell the fragrance of skunk on me. Her information was not only surprising, but embarrassing as well. I had showered extremely well and had nearly bathed in English Leather aftershave lotion.

If it had not been for my date I would not have realized how absorbent human hair is to scent. The scent my date was smelling was coming from my hair. Though I had

Odors within our clothing can be greatly reduced by washing with a oderless soap. Draper Mauldin's first throphy buck with dad.

washed my hair in a mild shampoo, she could still smell the aroma of skunk on me.

Hunters often fail to realize we carry odors to the woods—the odors created when we think we are decreasing them. This is the reason I recommend using soaps and shampoos which contain very low amounts of fragrances. Whenever possible, I use unscented soaps and shampoos. These products can be found at many sporting goods stores and are worth the money, in my opinion.

SCENT AND YOUR EQUIPMENT

There should be little question by now that I take scent very seriously. Some say I take it too seriously, and that's alright with me. The reason I take the subject so seriously is I take hunting deer seriously. The reason I take deer hunting seriously is I take eating venison even more seriously! In short, I try to leave nothing to chance. This is even true of the my hunting equipment.

Many hunters fail to realize how much odor we carry into the field with our equipment. I have seen some hunters be very conscious of their personal hygiene and never give their equipment a second thought. In many cases it is these same hunters crying "boo-hoo" because they were unsuccessful. So let's review a few elements which can help in eating more venison.

AVOIDING SMELLY APPAREL

When I began hunting deer my basic hunting apparel was whatever I had on. I gave little, if any, thought to the clothes I was wearing. It did not matter to me if I had just finished feeding the pigs or had just changed the oil in the tractor. I was ready to go hunting and time was of the essence. Today I have a totally different outlook on my hunting apparel.

I have learned the importance of keeping my clothing as clean and odor free as possible.

Unlike the days of my youth, hunters can prepare their hunting clothes more efficiently. Today there are products which will enable hunters to reduce scent from clothing. A product known as Sports-Wash is one I feel will greatly deplete odors from clothing. This is the reason I wash all my hunting clothes first before placing them in a "hunting environment."

The environment I refer to consists of plastic bags filled with natural materials. This is the reason I keep at least one hunting suit in a bag of cedar burrows at all times. I have another bag which is filled with freshly fallen oak leaves for hunting hardwoods.

The natural materials may consist of leaves, burrows or other scent-producing materials from the hunting area. The key is to be sure the scent is something which is natural to the hunting site.

I know there has been a lot said about the curiosity of deer and how strange odors will often attract them. This I have experienced numerous times over the years. I have seen some young deer react in a manner which I felt may have been due to their curiosity for the scent.

I have also watched numerous whitetails go bouncing away through the woods responding to the scent of pine in the middle of an oak grove. I simply believe the olfactory capabilities and the intelligence level of the deer are to be respected at all times. Deer know what belongs in the immediate area, *they live there*!

AVOID THE RED FLAGS

Previously I related the occurrence of the buck which smelled my scent on a small tree. I also related how I watched the buck turn away and disappear into the woods. This was another hunt which was foiled because I had not taken the time to think about what I was doing before I had done something.

Many hunters fail to realize how they place their scent without ever knowing it. Hunters should always remember that anytime we touch something we are placing our scent upon the object. This emplacement is due to the oils found on our skin. When we touch an object these oils are placed upon the object and can remain there for hours or even days. This is the reason I recommend wearing rubber gloves while traveling to and from hunting locations. With the aid of rubber gloves we can avoid leaving our scent on objects.

I have heard many hunters who say they avoid touching anything in their hunting area. This is good, but if they are not touching anything, how are they getting into their hunting stands? I know many hunters who hunt from permanent ladder stands. These stands remain in the woods all of the hunting season. Therefore, each time the hunter climbs the stand without gloves, scent is being placed on the stand. Think about it! How much of our scent will be on the stand after a few hunts?

STAMPING IN SCENT

There comes a time in every deer hunter's life when he locates a buck which keeps him awake at night. He spends many sleepless hours trying to figure out how he will har-

Hunters often fail to realize they are placing their scent on their hunting stands.

vest the buck of his dreams. I have had many bucks on my mind which resulted in wearing out the carpet. One such buck occurred when I was in my mid-teens.

Teenhood was a very special time in my life. During these years I was pretty loose and fancy free. The only main concern I had was fishing and bowhunting. Like most good bowhunters I had located an exceptionally nice buck days before the season would begin. Opening morning seemed like it would never arrive, but it did.

The sun was barely above the horizon when I spotted my trophy buck coming through the woods. The deer was walking straight towards me and my legs were beginning shake. My nervousness meant that the buck was getting close and I was preparing for the shot.

The deer never slowed down until it was approximately 50 yards from my treestand. This distance, I felt, was out of range for my old recurve bow. All I could do was stand

The nose of a deer can easily detect odors carried by a hunter's footwear.

there and watch my dream buck sniff the ground for several long agonizing seconds.

Slowly the deer smelled the ground as it took two or three steps towards me. With these steps my heart began beating faster as I thought the buck was again on its way. It was perhaps with the third step the buck took that it suddenly wheeled around and bolted the way from which it had come.

Long minutes passed by as I stood in amazement and tried to figure out what had transpired. Knowing I had not made any sound and the animal had not seen me, the shock was quickly replaced with knowledge. I knew the buck had smelled me. But wait, the wind was coming from the deer and towards me! These thoughts brought forth a few minutes of bewilderment before I realized what had happened.

Replaying the encounter in my memory I remembered what had transpired. I recalled the way I had approached my stand that morning. I had walked in the direction where the buck had stopped. Was it possible the deer had smelled my footsteps?

After some deep thought, I realized I was wearing the same boots I wore everyday. The same boots which had been exposed to everything I had been in contact with. The same boots I had worn while filling my car with gasoline. The same boots which I had worn while crossing the pig-pen to go to the woods. It was with these thoughts I began respecting the olfactory capabilities of the whitetail even more.

The remainder of that morning was spent with me going to town and purchasing a pair of rubber boots. With the boots at home they received a good washing with detergent and water. After the washing the boots were placed in a storage shed.

Those boots, like the many other boots I have had since, never were worn except for hunting. Those boots, like all the others, received an abundant smearing of dirt before going to the hunting stand. Those boots, just like the others since, never alerted a deer to my presence again. This is the reason you will never find this ol' boy hunting in nothing but rubber or rubber soled boots.

THE UNTHINKABLE

There are hunters and there are hunters. Some hunters consider the obvious, some go beyond. Those who strive beyond the ordinary are usually those the obvious become jealous of. The hunters who go beyond know that, in most cases, it will be the little things which determine the big payoffs. I refer to payoffs as the big trophy bucks all hunters dream of.

The little things I am referring to are little things like excess lubricants on some of their equipment. Many times I have seen hunters use way too much oil on their guns.

This oil, my friends, contains a foreign odor to any deer which might browse around your hunting area. It is also this same oil which can get on your hands while the gun is being carried. It will contaminate the cleanest of rubber gloves. So when oil or other lubricants is being used on equipment, use it sparingly. I also suggest just wiping all excess lubricants from the article before taking it afield. When in the field, avoid touching the areas of the gun which contain the lubricants. Remember, it is these things which add up!

CHAPTER FOUR

THE COMMERCIAL SCENT INDUSTRY AND THE HUNTER

I have been affiliated with the commercial scent industry in various capacities. My duties have included everything from researching products, product development, and marketing.

I have lectured all across North America on hunting with commercial scents. The reason for much of my involvement in the commercial scent industry is I have been successful at hunting with these scents. In short, I will not go deer hunting without them for the reasons we are about explore.

I can still remember the first bottle of commercial deer lure I ever bought. I was very excited about possessing this so-called magic potion. Somehow I felt the content of this bottle was the answer to all my deer hunting problems. It was the ignorance of youth which made me believe success was in that bottle. I was committing the same mistake as thousands of hunters have done! I was relying on the commercial scent to make me a better hunter.

Seasoned hunters know there is nothing that can replace woodsmanship. Woodsmanship, in my opinion, is the level or degree to which a hunter has been educated, and this is usually from actual field experience. Experience is gained from a hunter learning from his or her mistakes, and mistakes are common to people who do not think or respect the animals they hunt.

THE MAGIC BOTTLE

The very first time I tried using a commercial scent product is a cherished memory. I remember it and wonder how I could have been so naive. I sometimes ask myself, if I was that gullible, it is a miracle that I ever shot a deer. I also believe if one cannot laugh at one's self, one has no right to laugh at anyone else! Therefore, I hope you have a hearty laugh on me!

HOW IT ALL STARTED

The local sporting goods store in my hometown was somewhat of a shrine to me as a lad. On the walls of the store were the shiny new guns of which I dreamed. As it turned out, one of those guns would be given to the hunter who checked in the biggest buck of the season. I was determined to be that hunter.

The pegboards in the store were filled with all kinds of hunting accessories. I just knew these accessories would make me a better hunter. The store's owner was an old gentleman whom I thought was the world's leading authority on deer hunting. Whatever this man said was pure gospel to my ears, that simple! Today, however, I know that gentleman was the slickest salesman I have ever met. In most instances, he had pocketed what little money I earned before I ever got it!

This holds especially true when what is known as today's commercial scent was revolutionized.The first bottle of

Many years have passed since the author began using commercial scents. This is only one of many fine bucks he has taken while using a commercial lure.

Today the commercial scent industry has a vast assortment of products to select from.

commercial lure I ever bought was a bottle of Davey Bracken's deer lure. The best I can remember, the bottle showed a picture of a hunter running from a big buck and shouting, "Help, Davey, Help!" It was because of this advertising and a little persuasion from the store owner that I was hooked. In fact, I was so sure of myself and my new secret weapon, I bought an abundant supply of shotgun slugs. With my newly purchased equipment and high hopes, I headed for home.

At home that night I could hardly sleep from worrying about the following morning. I simply could not decide if I wanted to shoot the big 10-pointer in the north pasture, or the 12-pointer that stayed around the eastern cornfield.

The night was also spent with me visualizing one of the bucks charging the spot where I would place my new scent. I could visualize the buck pawing and scraping the ground only moments before I pulled the trigger.

The next morning arrived with me traveling in the dark to our eastern cornfield. This was the morning I would become the county's top deer hunter. There was no question in my mind that my secret deer lure would help me win the new deer gun.

Today I laugh as I recall those memories and my disappointment of that first day. In fact, I think my fatigue was greater than my disappointment that day. The only deer I saw awoke me when it began running from where I had placed my secret weapon. My secret weapon, as I learned, was not a weapon, but a *tool*.

TOOLS OF THE TRADE

Decades have passed since that day and many lessons have been learned regarding commercial lures. I would also like to mention that scores of deer have fallen prey to me while learning these lessons. I believe I could not have learned these lessons if I had not kept an open mind to the subject of scents, and if not for the development of what is now the commercial scent industry.

It has been during the past decade or so that the commercial scent industry has grown to its present day status. It has been with this expansion that many new products have been developed. I feel these developments have aided thousands of hunters in becoming successful. Having met and talked with many of these hunters, I feel safe to say we agree many scent products are merely tools which increase our odds when used properly by a knowledgeable hunter.

These same tools will decrease our odds for success when not used in a proper manner. In summary, a bottle of commercial scent is no different than a doctor's scalpel. If the scalpel is used by a skillful doctor, a positive result can be expected and vice-versa.

COMMUNICATING WITH COMMERCIAL SCENTS

I have become a firm believer that certain applications of commercial deer scents do work. I feel there are certain aspects, or elements, which will govern the outcome of using a commercial scent.

One such aspect is that deer communicate with scent more than any other form of communication. Although we reviewed this in Chapter Two, there are certain elements that I feel need to be covered at this time.

One such element is deer sometimes do not wish to communicate with other deer. It is my opinion from observing deer that they, like humans, have their own personalities. Some deer are simply more social than others. I firmly believe this applies to the dominant animals of the herd.

There have been numerous times I have watched captive dominant bucks lay in their beds all day. During this time the other captive deer stayed away from the bedding buck. It appeared as though the other animals could sense the buck wanted to be left alone. Surprisingly, I have witnessed this while sub-dominant bucks chased estrus does around the pen. The dominant buck's reaction to this was, "If you want her, you can have her!" But in reality, I think the dominant buck knew the doe would not submit to the lesser buck.

The point I am making is, the dominant buck simply did not wish to communicate. It has been from observations of this nature that I have formed some opinions about commercial scent products and deer behavior.

The author believes commercial scents are tools, just like camouflage and a good bow. If used in a correct manner, they can increase the odds for successful hunting. (Photo by Judd Cooney)

CONFUSION WITH COMMERCIAL SCENTS

One of the most often asked questions is, why do deer often pay little or no attention to a commercial scent? To answer this question my first reply, as I have already stated, is they simply did not wish to communicate. The second best reason I think is the hunter represented a confusing situation to the deer with the scent.

I think that few humans, if any, have the expertise to know what a deer is thinking. I know I certainly cannot, even after dealing with them for over three decades. I feel the best anyone can do is simply develop a hypothesis — an educated guess based on what the animal had done and from past experiences. This, I feel, might answer why a deer reacted as it did.

I have seen reactions from deer as if they were confused on several occasions. Most of the time this occurred while I was experimenting with various scent mixtures. It has been from these experiments that I have formed some very strong opinions about mixing scents.

I believe hunters often take for granted that we are relaying a strange message to the animals with the scent. I also believe this is the reason some hunters experience some of the negative results related to commercial scent products.

MY FRIEND, THE CHEMIST

A few years ago I hunted with a friend who thought he knew everything there was to know about deer hunting. There was no question he was a good hunter and took a lot of deer. The only problem was he knew all the answers before the questions ever arrived. Somehow I knew my friend's supernatural knowledge of deer behavior would halt him someday, and it did!

The sun was shining brilliantly one fine October morning. For hours I had sat in my deer stand over looking a large soybean field. Across the field my friend guarded the section where the field joined a cedar grove. There were a

The author believes deer have their own personalities and are often loners.

lot of deer signs present at my friend's location and I knew his chances were good for encountering a nice buck. It was with this thought in mind that I used my binoculars to frequently spy on my friend.

While doing one of my periodic checks I noticed some movement in the cedars. The movement was approximately 20 yards slightly to the right of my partner's stand site. I focused the binoculars there and discovered a large set of antlers partially hidden in the cedars.

The antlers were now receiving my undivided attention as I watched them move up and down. The movement of the antlers was now captivating me as I could now understand what was taking place. I felt something strange was happening as the buck would only move its head upward and downward.

Several minutes had expired since I had begun watching the buck before I glanced at my partner. Though he was

The author advise to never mix various types of scents if success is expected.

well camouflaged and fairly hidden, I could see that he was facing the buck. I could also see that numerous small cedar trees were in his line of fire. There was no way possible for my companion to attempt a shot at this point in time.

The show lasted for approximately another two or three minutes before the buck slowly turned and vanished into the thick cedars. Soon after the buck disappeared I noticed my friend had retired from his treestand and was coming my way. With my hunt now foiled all I could do was hopefully meet him before he stumbled around my hunting site.

Hurriedly I joined my companion in the center of the field. His language was quite strong as he began cursing a couple of commercial scent companies. The companies were the reason the buck did not present itself to him. He was sure the product he had bought was bad.

Knowing my friend as I did I listened to his remarks for several minutes. Then when he took time to breath I asked him what lure he was referring to. His answer somewhat startled me as he named three different major scent companies.

Before the last word of the last company name left my friend's mouth I said "*What*! What do you mean three different brands of scent?"

Then before he could say another word I said, "You mixed three different types of lure to form one? Are you crazy?"

Realizing the disappointment of my friend at this time I voted for breakfast. My deer know-it-all buddy was feeling quite low. I thought I would wait awhile before explaining the mistake he had committed. The mistake many hunters make is to blame the commercial scent for the results.

THE BIG DIFFERENCE

In the case of my friend he had combined three different types of scents to create one. The reason he had done this was because he had a small amount in each of the containers. He simply thought it would be better to combine the three into one bottle for convenience. He paid the price!

Hunters should always remember that most scent companies produce their own products. This means there will be a difference in the scent of each product due to various factors regarding the animals which are used to produce the product. This is very true of the assorted types of urine-based lures.

Animals which are used to produce urine lures may exist in different geographical locations. This means the diet of the animals could be very different. It also means that difference can be transmitted though their urine deposits. Combining various urine can and will create an unnatural scent. This is the reason I recommend using a single brand of urine scent at all times.

MY THEORY ABOUT URINE LURES

In the minds of some people, the commercial scent industry is nothing more than a fraternity of con artists. I freely admit I have encountered some people who acted pretty suspiciously. I will also admit many years have passed, and these encounters have not transpired during the past decade.

I will further admit I feel the commercial scent industry is pretty darn lax in educating in the ways of using their products. This, my friends, is nothing new in that within most industries, marketing comes first and education comes second.

This is the reason I am writing this book and traveling the country lecturing on the subject—to simply share what I have learned about using commercial scents.

While I am on the subject of scent usage I would like to pass along something which I feel is very important about scent usage, and something I think many hunters neglect and few companies mention.

THE ATOMIC BLAST

It is my opinion many hunters fail to understand what commercial urine lures are. They are the urine of a deer which has been placed into some form of container. So what is so important about that? Plenty, if a hunter doesn't

The olfactory capabilities of a deer are many times greater than a human's. Hunters should take this into consideration when applying commercial scents.

think about some important factors which will make the difference in how successful we will be in the woods. Why? Because I think many hunters do not understand some of the elements essential to deer urine.

During warm weather conditions, the author dilutes his lures before using them. Daniel Barfield experienced success using this theory.

Urine is high in ammonia and when stored will become even stronger smelling with time. We should know this by removing the top from our lure and taking a good sniff. In most cases the human nose has no problem detecting the scent.

Now, if the human nose can detect the scent, how do you think the nose of a deer will detect it? Common sense should tell us the deer will receive the scent message many times greater than the human.

For years, various scent producers instructed hunters to use the concentrated lure in the same manner we would use fresh urine. Fresh urine does not have the strong smelling odor of the concentrate! Therefore, when applying the concentrate, are we not creating an unnatural scent to the animal? I think so! I have seen negative reactions when using the full strength lures. This is the reason I dilute the concentrate with spring water. By doing this, a more natural smell will develop.

The amount of dilution will depend on a couple factors, the first being the original amount of odor from the commercial scent.

The amount of odor will vary with the shelf time of the lure.

The other factor is the existing hunting conditions. This is due to how some weather conditions can and will effect the olfactory capabilities of the deer.

During periods of warm and humid conditions I may dilute the commercial lure by as much as 50 percent. This is due to the animals being better able to smell. Cold and dry conditions work the opposite on deer. During these conditions the amount of dilution will be reduced, if at all. In short, learning to dilute the lure comes from experience. The more experience a hunter has, the more efficient he/she will become.

CHAPTER FIVE

ARE COVER SCENTS IMPORTANT?

Cover scents are, in my opinion, one of the most misunderstood products found on the scent market. The reason I make this statement is that most people believe these products will eliminate human scent. This myth has been developed by some fancy advertising.

Cover scents simply do what they say they will do, to a degree, and that is they will distort or camouflage human scent. I firmly believe there is nothing on the market which can completely eliminate human odor for long periods of time.

I do believe we can camouflage our scent and confuse the animals with cover scents. This is the best I feel anyone can hope for when using any cover scent product.

Cover scents are the earliest form of scent known to man. It has been said the American Indians used cover scents. This would come as no surprise as the Indians depended on hunting for survival. These hunters used whatever means they could in order to be successful at their skills.

The first encounter I had with cover scents came from reading about the Indians. As a youngster I read everything possible that I thought would make me a better hunter. Through these readings I gained quite an education in using cover scents.

OUT FOXING THE FOX

Like many young hunters the words of valuable hunting information were hidden within my text books. Often as not, this valuable information was cut from a hunting magazine and taped to the pages of my English book. With these alterations I garnered the data and applied it when I got home.

One day while the rest of the classroom learned something (I still don't know what), I began an education in cover scents. From the treated pages of my book I read how the Indians would use animal urine to cover their human scent. This, I felt at the time, was a major breakthrough in my hunting career. The remainder of the day was spent with me visualizing getting within grabbing distance of a big buck.

The ride on the school bus that day was the longest one I could ever remember. I simply could not wait to get home, change clothes, and pour fox urine all over me.

Like the Indians I would the slip up and arrow me a big buck before dark. Never did I think I was pouring half my winter's supply of trapping lure away! Never did I think about the consequences of what I was doing.

Later that afternoon I watched the sun disappear as I went home empty-handed. The two doe deer I had encountered were probably still laughing at the goofy-looking fox they had winded. In short, I knew I had more to learn about cover scents.

I also learned that my grandmother did not exactly relish the fragrance of fox pee in the house!

The following day was spent with all my school friends discovering a fact about nature. The lesson they learned was there were more aromas in the wild kingdom than just

This trophy buck fell victim to the use of Acorn cover scent.

those of wild flowers. Since we were in a place of education, I felt the day had been well spent.

The return trip home was much like the previous day, slow. And, like the day before, I soon was in the woods with my bow and arrows. This time I would allow the remainder of my trapping lure to cover my scent from a tree stand.

The hunt was going as usual with the sound of feeding squirrels pestering me. With each rattling sound of the dried leaves I knew my trophy was there, only to discover one of the little rodents.

Daylight was fading fast when I heard a disturbance in the leaves behind me. At first I thought it was late feeding squirrel until the sound of a breaking twig reached my ears. Knowing that whatever had broken the twig weighed more than a squirrel, I slowly turned around.

The shock of seeing the large rack of the buck sent my heart to beating rapidly. I could sense that within moments the buck would be within easy range of my arrow. In mere seconds I would join the ranks of other master bow hunters. I just knew there was no way this buck could escape me.

My confidence increased when the deer suddenly stopped and stood motionless. The buck's action was my cue to prepare for the shot. I was in the process of doing just this when the buck suddenly stretched its head in a forward motion.

The forward movement of the deer placed the animal's nose far beyond its body. It was also with this extension that the buck began raising its nose upward. The movement of the animal resembled some form of a tracking device, which was quickly locking in on my exact position.

Realizing something was not kosher I began drawing the bow as the eyes of the hunter and the hunted met. This would not be the final hunting mistake I would make that afternoon. The final mistake of the day was when grandma smelled me enter the house!

LEARNING FROM OUR MISTAKES

The encounter with the buck replayed in my head for much of the night. The memory of how the buck had zeroed in on me was simply amazing to me.

I asked myself over and over how the animal had known I was above it. I felt that if the deer smelled me, a human, it would have been spooked! I also believed there was very little chance the buck could have detected me with all the fox urine I was sporting. With this last thought, I began theorizing what had happened.

There was no question in my mind that the deer had smelled something. There also no question in my mind that

The use of a cover scent is very important for hunters demanding close range encounters with game.

The author believes the intelligence level of a deer should be respected when applying a cover scent. (Photo by Judd Cooney)

the deer had detected the scent of a fox. The only question was, was the deer smart enough to know that red foxes don't live in trees? I decided I needed to unlock some mysteries pertaining to the intelligence level of whitetail deer. This would prove to be quite a feat for a 14-year-old boy.

Application of scents should be part of an overall hunting strategy, just like method of hunt and placement of stand. (Photo by Judd Cooney)

RESEARCH IS REASONING

On my grandma's farm was a fairly large field of approximately 16 acres. This particular year the field had been planted in corn and harvested. Deer were very abundant about the field, therefore my research center was born.

I place several treestands around my private research center. I then placed on the treestands clean pieces of cloth that had been soaked in various ingredients. These ingredients consisted of typical aromas which were then commonly used as cover scents. Skunk and fox urine were among the top leaders at that time. So it was these and some natural scents such as pine-scented cleaning products I used the most.

Much of that hunting season was spent with me observing the reactions of the deer to the various odors.

Most of my findings were that the deer would become very nervous to a scent which created an unnatural situation, such as the fox scent being in a tree!

On a few occasions I watched the deer become curious to a scent and cautiously move towards it. These actions usually resulted in the animal's caution overpowering its curiosity. From these observations I have learned to respect the whitetail deer even more than I previously had.

Respecting the intelligence level of the whitetail deer has become my top priority, which is now the basis for all my hunting, and hunting strategies. I believe respecting the animal's intelligence and senses are the key factors in being a successful hunter.

Today, unlike the days of my youth, there are many products which will enable hunters to become more successful. Success which comes not only from respecting the animal's intelligence, but from respecting our own, as well! This is why I think cover scents are very important for the serious deer hunter. In fact, so important I don't leave home without them!

CHAPTER SIX

THE COVER SCENTS

In recent years the commercial scent industry has helped hunters smell like almost anything imaginable. Today a hunter can smell like a tree or an animal. This can be accomplished by simply removing a bottle top and applying a commercial cover scent.

Unfortunately I feel there are a few factors which many hunters fail to think about before dousing themselves in these products. I feel many hunters fail to think about what we covered in Chapter Five—the animal's intelligence level! I also think so many hunters fail to believe a deer's nose is as powerful as it is.

Some hunters simply believe staying downwind of the deer's direction of travel is the key. There is no arguing that being downwind is important, but how can we always be sure the buck of our dreams will not approach us from another direction?

It is my opinion that the hunter who is always prepared is the successful hunter. This is why I value the use of cover scent so much. Cover scents simply increase my odds for hunting success!

IT IS NOT AS BAD AS IT MAY SEEM

Some hunters do not like to use cover scents because of the foul smelling odor some cover scents produce. I know

The author feels hunting success comes from never underestimating the nose of a deer!

this for a fact as I have spent hours with the essence of skunk on me.

Today hunters can apply a cover scent without gagging from the effluvium. Hunters can now find cover scents with the fragrance of pine, cedar, dirt or various other scents. Most of these cover scents are in the form of a thick oil. I refer to this type of scent as *oil base* scents.

In many cases I recommend these products over other cover scents, simply because, as previously mentioned, some people cannot tolerate strong pungent odors. I also recommend oil base scents for those who have family members with a relatively good sense of smell.

I often use oil base cover scents during the warm days of archery season. The reason for this is because of insects.

Insects such as mosquitoes, black flies and other pesky, biting insects can drive a hunter crazy. When these lovers of human flesh become a problem I combine a high potency insect repellent with an oil base cover scent.

The odor of the insect repellent is weakened or absorbed by the oil base cover scent. This has often aided me in controlling movement while on stand. Unfortunately, this may not work for everyone as it can cause skin irritation on some skin types. This is the reason I suggest mixing a weak mixture of approximately two-thirds cover scent to one-third insect repellent and test it on a small area of skin for any irritation which may occur.

KEEP IT NATURAL

I cannot stress the importance of keeping a cover scent natural to the existing conditions. When using an oil base cover scent, it should represent something that the deer are familiar with in the area. Often hunters try to mix the scent of pine trees in a hardwood forest.

In my opinion this is no different than a cook blending chocolate syrup in spaghetti sauce. The aroma is very easy to detect.

OTHER USES FOR OIL BASE COVER SCENTS

There is no question in my mind that many of my friends make fun of me. They feel I am way too scent-conscious because of things I do. In many cases it is these same people who accompany me while hunting. The reason I hunt with these people is because I often need help dragging my deer from the woods. So allow me to share with you some things I don't share with some of my irate friends.

One of the things which I feel save a lot of deer each hunting season is noise. Low volume squeaks created by our hunting equipment is usually due to the lack of lubrication.

This is why I often use an oil base cover to lubricate various parts of my bow. I simply replace common lubricants with the cover scent. This way I am killing two birds with one stone. I am decreasing noise, and desensitizing a nose!

THE OTHERS

Many hunters want to leave nothing to chance while hunting. It is usually these same hunters who will apply a urine-based cover scent. The reason for this is, urine-based cover scents are much more potent in odor.

With the increase in odor, a hunter should expect better results in disguising their human scent. This is the reason I use urine cover scents most of the time. The trick is to use a scent which will not alarm the deer and use it in a natural state.

If the scent of a skunk, raccoon or fox is used properly, it will represent a natural state to the deer. A natural state to a deer is a state of safety.

THE USE OF FOX URINE

I relayed my early experiences of using fox urine as a cover scent in a previous chapter. Though it may have appeared I refrain from the use of fox, it's quite the contrary. Today I still use fox urine for a cover scent when the hunting conditions call for it. The conditions I refer to are hunting from ground level locations.

In fact, I believe the use of fox urine helped me harvest one of my all-time big bucks. I also believe the experience I had with the buck taught me a valuable lesson pertaining to wildlife. The lesson is that some species of wildlife help

The author checks his equipment frequently for noise. Oil base cover scents are his preferred lubricant.

to wildlife. The lesson is that some species of wildlife help other species during times of stress.

DEALING WITH A DILEMMA

Like many hunters I have spent countless hours hunting on public hunting lands. Like some hunters of public lands I have also encountered some not-so-smart hunters on these lands. Hunters who had little, if any, regard for their hunting comrades. Hunters who would shout at the top of lungs to locate their partner. Hunters who would wander around the woods all day trying to "Sneak up on a deer." Hunters who would sit in their treestands drinking coffee, smoking cigars and listening to the football game. Yes, I have encountered people listening to a transistor radio while hunting.

Experienced hunters quickly discover deer seem to disappear after the first day of season. These same hunters also know the deer are simply avoiding the areas of human involvement. It was with this I applied a plan for hunting public lands.

WHO OUT-FOXED WHO

I had been hunting a certain section of some public hunting land for two days. Like many public lands, each hunter was assigned to a specific location. I had been assigned to a location at the base of a long ridge. I knew someone else had been assigned to a place on the other end of the ridge. I knew this from his communications (shouting) he had with his friend on the other ridge. I also knew a buck had created numerous scrapes along the top of the

Crafty hunters like Harold Knight know how to deal with hunter-wise bucks on public hunting lands.

ridge. From the stand my dear friend was hunting he could easily shoot the buck.

During the time I had spent hunting my location, I had seen four or five does. All of the deer had traveled between our stand sites. With this information I was sure the basic travel pattern of the deer was confirmed. The only problem was that the undergrowth was quite thick and I might not see the buck before the other guy did. I knew I had to do something to detour the buck in my direction. I had to do something which would alarm the buck, but would not spook it. I knew there was no chance of me getting a good shot through the thick undergrowth.

The author believes the scent of threatening animals should never be used while hunting.

The following morning found me in the woods approximately 30 minutes early. I had checked the wind direction and knew if the buck ever reached the ridge top it could smell me. I also knew if the buck reached the top the other hunter would have a shot. I figured I would have to alarm the buck about halfway up the ridge. If I could alarm the buck at this juncture I felt it would turn and travel into the wind. This would have the buck approaching me in a cross-wind direction. This, I felt, could be accomplished by re-creating the mistake I have previously mentioned.

The extra time I had that morning was spent bending small trees over and placing fox urine in their tops. I placed the fox urine on five or six trees and headed to my stand. Now all I could do was hope and wait.

The morning began as the previous two mornings had. The sun came and a few shots echoed along the countryside. I had watched the other hunter enter his tree-stand with the aid of a flashlight, so I knew someone was there.

The sun seemed to rise slowly that morning. I was wanting the sun's warming rays to come quickly as I was a bit cold. It was during these moments of wishing for warmth that I noticed some movement coming through the undergrowth. The movement was quickly identified as the large rack of the buck was clearly visible.

To this day I do not know if it was the scent of the elevated fox urine which sent the buck my way or what. The only thing I know for sure was that the big 10-pointer dropped at the crack of my rifle!

Today in many parts of the country fox urine is still a very popular cover scent. I often use it when the hunting situation calls for me to hunt from ground level. Therefore, I feel we have reviewed the importance of keeping everything natural to the nose of the deer.

OH, THE ESSENCE OF SKUNK

I can only wish I had a nickel for every hour I have sat in a deer stand gagging on skunk scent. To my knowledge, skunk scent was one of the first commercial fragrances we had for a cover scent. Today I use skunk scent occasionally to divert deer to me as I have used fox scent.

To have skunk scent near my exact hunting site is strictly a no-no in my book. The reason I feel this way is because of what I think skunk scent represents to a deer and why.

During the past few decades many portions of the country have been invaded by the coyote. Another type of canine which is becoming more and more plentiful is the feral dog. Deer, for the most part, will do anything possible to avoid encountering these canines. These are canines which often will go out of their way to kill the common skunk.

Skunks use their pungent odor to ward off their attackers as their first line of defense. It is from this natural form

The author feels the scent of the raccoon has helped him harvest many of his deer to date.

of defense that I believe deer often relate the scent of a skunk to possible danger.

In all my years of deer hunting I have never witnessed a deer fleeing from a skunk. I have, however, seen deer being chased and killed by coyotes and dogs. Think about it!

UP OR DOWN RACCOONS ARE AROUND

In Chapter Five I related my experiments with a lot of various scents. Scents that, during that time frame, were not on the commercial market, but ones I used for many years because of my trapping background. These scents have aided me in harvesting scores of deer, which may not have been possible without them.

The cover scent which I have come to depend on the most is raccoon urine. I began using raccoon urine during my mid-teens after applying some thought to the subject of cover scents.

During my early days of scent exploration I sought for a scent which would provide two important factors. The first factor was to provide an odor strong enough to camouflage mine. The second was a scent which would be natural to the deer at any state, and raccoon fit the bill for both factors!

My first thoughts of using raccoon urine for a cover scent came early one morning. Today I still remember watching those two raccoons cross the woods and climb up their den tree. When the raccoons had climbed about halfway up the tree, they nested within a fork of the tree. I figured the warming sun felt as good to them as it felt to me. I remember watching the raccoons until a big whitetail doe suddenly appeared. The deer slowly began feeding around the base of the same oak tree the raccoons called home.

What I remember most about the doe was she never once raised her head to check the air currents. The deer's actions were somewhat surprising to me. I could not relate to any deer which had been so relaxed for that long a period of time.

Time moved along as the deer fed on the freshly fallen acorns of the red oak tree. During the half hour or so the doe fed, two other does joined her in feeding. These deer reacted in almost the same manner, total relaxation.

Being the inquisitive fellow I am, I began thinking how the deer could not help but smell the raccoons.

It was with these thoughts that I realized deer had nothing to fear from raccoons. They had no reason to pay any attention to the scent from above. I also remembered the raccoon urine I had ordered for training my raccoon dogs.

MY SECRET SCENT

Anticipation was flowing within me like a river the next morning. I could hardly wait to cover myself in raccoon pee-pee and head for the woods.

The walk from the house to one of my treestands was accomplished quickly and I was ready to hunt. A hunt which would prove whether or not my theory was valid.

The validation came within the first hour I had been in the stand. I had detected an average size six-point buck slowly browsing to my right side. The buck was traveling in such a direction that I figured it would come into contact with my scent. I also figured the buck would be in range of my arrow at about the same time as it would smell me.

The buck soon proved my calculations to be pretty accurate as it stopped suddenly. The buck stopped almost at the exact position I had predicted it would. The surprising thing was the buck sniffed the air currents briefly before resuming its browsing.

Long moments ticked away as I allowed the deer to decrease the range between us. During this time the deer raised its head a few times to sniff the air. At no point in time did the deer ever become nervous or suspicious of my presence.

It was not until the deer had lessened the range to 15 yards and felt the broadhead strike him that it spooked.

That buck was the first of many which encountered the effects of using raccoon urine as a cover scent!

CHAPTER SEVEN

APPLYING COVER SCENTS

I have often been amazed by the way some of my hunting friends suddenly go brain dead. This usually occurs within moments after we stop the truck and head for our tree-stands. It is as though a spell has been cast upon them and they simply cannot reason anymore. This is especially true when it comes to applying their cover scents. My dear friends will simply not take the time to apply a little common sense before they do something. A very good example of this happened one morning a few years ago.

After a long ride to the place where we were going to be hunting, we finally got out of our truck. The instant we did this, my friend grabbed his bottle of cover scent and began pouring it on his boots. Being somewhat amazed at his actions all I could do was stand and watch.

I watched my friend empty an entire bottle of premium brand cover scent over his rubber boots. With the last drop of the scent being drained from the bottle he remarked, "Now I'm ready!" With this I could not refrain myself from asking, "What are you ready for?" His reply was that he was "ready to go hunting now that I have my cover scent on."

In the dim light of his flashlight I could see my friend was serious about his hunting. I could also tell he was becoming frustrated with my teasing. I then decided it was time to administer some advice to ol' boy as we walked.

"Mark, how effective is that cover scent going to be for

you?" His reply was "Plenty, why?" I then informed him that 99.9 percent of it was soaking into the ground behind us. I also informed him that he wasn't planning to stand in that one spot all day where he had wasted the scent. The information was instantly received by my buddy and the result was a sudden halt. "Damn, how dumb can I be?" he said.

It has often amazed me how some hunters commit the mistakes they make when applying cover scents. Often these mistakes are made by simply being in a hurry to get into the woods. Like my friend they simply fail to think about what they are doing at the time. I also feel some mistakes are made by improper instructions within the packaging.

I have read some instructions which have accompanied the product and frowned to myself. The opinion I formed after reading their words was the author was more concerned in selling the product than in hunter success. I also think some people in the scent business should take up the sport of deer hunting before dispensing advice on the subject. These are simply my opinions, nothing more, nothing less!

ENCOUNTERING AN EDUCATION

Like most youngsters I thought much of my school learning was a waste of time. I could not understand how some of the class subjects would influence my life forever. Today I know, but yesterday is gone! Anyway, a valuable lesson pertaining to cover scents was relayed one day in the classroom. Though I doubt the teacher realized he was conducting a form of hunting seminar, he was to me.

The subject of the lecture was about molecules and their

Proper cover scent application is especially important for the bowhunter.

weight. The instructor informed us how temperature effected the weight of the molecule. The warmer the molecule was, the less weight it had. The less weight it had, the higher it would rise until it began cooling. When the cooling process began, the molecule would begin to become heavier and then drift downward. He illustrated this process with something which produced some red smoke.

While my classmates ooohed and ahhhed at the smoke, I began thinking about scent molecules and deer hunting. I then started thinking about the scent molecules escaping from our bodies. These molecules should have the same temperature as contained in our bodies. In most cases this would hold true at a temperature of 98.6 degrees fahrenheit.

I then realized that the colder the temperature was, the quicker the scent molecules would cool and drift downward. In many deer hunting situations this would occur quite quickly.

This, I thought, would result in the animal receiving our scent at a lesser range, or would it?

Applying some thought to the subject I then realized the velocity of the air currents would influence the range, as well. Now I would be hunting deer armed with scientific data. Now I would be able to calculate where and when a deer could smell me. Things were definitely looking up for me, I thought.

That afternoon I went to my favorite deer stand behind my house. At the stand I decided to act on my theory. It was not long after I had entered the stand that a spike buck proved me to be correct. The deer reacted to my scent almost at the location I thought it would. It was at this time I realized an important factor in controlling my scent.

If my scent molecules were rising above me, why was I placing my cover scent on the lower portions of my body?

Hunters often fail to realize scent molecules will travel further when dispensed from an elevated position.

It was with this thought that I began to wonder how these warm molecules were escaping from my body when I was covered in two layers of heavy clothing, plus a coat.

Now with my brain working in high gear, I began trying to put the puzzle together. During these moments of deep thought I heard my grandmother call me from the house. Turning towards the house, the answer to my scent question was answered!

At the first sight of the smoke escaping from our chimney, I knew. My scent was escaping from the primary opening of my clothing, the collar.

The collar of my coat presented the same escape route as the chimney did for our wood stove. It was with this information that I began applying my theory of controlling my scent.

Realizing how my scent was being transmitted, I knew I had to place my cover scent above me. By placing the cover scent above me, my scent molecules would than blend

The author believes periodic applications of a scent eliminator will increase his odds for success.

with the molecules of the cover scent. This would create a blend of scent molecules which would hopefully confuse an intercepting deer.

Today many years have passed and my childhood theory has become the foundation for the manner in which I apply cover scents. I have also incorporated some of modern technology with my methods, which has helped me in disguising my scent from the animals I hunt.

REDUCING CAN BE INCREASING

Unlike the days of my youth the modern hunter has many helpful products which will reduce our scent to some degree. These products are called *scent eliminators*. Having used these products I do believe they do help in reducing our human odors. This is the reason I never leave home without my scent eliminator.

Most human scent emanates from the collar and hair.

Experience has taught me that whatever method of scent covering I use, I always begin with a scent eliminator.

By using the scent eliminator first I am reducing the basic molecular structure of a scent molecule. This is done by the scent eliminator inhibiting bacterial enzymes which produce odor. This, in turn, will allow the molecules of the cover scent to become even more effective.

I will also wipe my hands and neck every so often while on the hunt for added protection. During most conditions this will be every hour or so. Therefore, we can see how a

The use of artificial leaves for applying cover scent is one of the author's preferred methods.

decrease can create a major increase in the odds for successful hunting.

APPLYING THE COVER

Throughout the years I have used various methods for dispensing my cover scent. Some of these methods may appear to be going to extremes for some folks. Some of the methods may seem to be not enough in the eyes of other people. In most cases I found that the existing weather conditions will be the major factor in determining the degree of application. In short, three important weather conditions will determine the application. The conditions I refer to are temperature, wind velocity, and humidity.

Hunters must remember their scent molecules will rise in warmer temperatures. These molecules will carry further with the velocity of the air currents.

Hunters should also remember the animals can smell better in humid conditions due to the increase of moisture in the air. Therefore, I base my cover scent strategy on these three important factors.

When hunting in warm, windy and humid conditions I enforce the highest level of cover scent application I know. The reason for this is we, the hunters, will produce more odor in the form of perspiration; therefore, these types of conditions will receive the umbrella method.

The umbrella method is simply what it says it is. I place myself under some form of an umbrella. In most instances the umbrella will be a small leafy tree limb. I will saturate the leaves with the cover scent I have selected. Many hunters are finding some of the new deer stand umbrellas to be very useful also.

If a small leafy limb is not available, I will use artificial leaves. Hunters can purchase small clusters of synthetic leaves at most sporting goods stores. If the leaves cannot be found in the sporting goods store, try a hobby shop.

Once I have the artificial leaves I place them in a zip-lock bag. The zip-lock will permit the fake leaves to lightly soak in the selected cover scent.

The scented leaves can then be easily attached to my cap and on the shoulders of my jacket. After hunting, the leaves can be removed and placed back in the zip-lock for convenience.

Another method of applying cover scent is with the use of 35mm film canisters. This is done by simply filling the canisters with cotton and to soak in the cover scent. The canister can then be sealed with its lid.

At the hunting site, the canisters can be re-opened and hung above the hunter with the aid of a clothespin or some other attaching device.

There are many different means of applying cover scent to the hunting site. Some of these ways will be described in the upcoming chapters. The most important factor of applying cover scent is to have the application above the hunter as we have reviewed.

CHAPTER EIGHT

THE ATTRACTANTS

The very first bottle of deer lure I ever saw had the words "ESTRUS DOE" written upon it. In my youthful mind I did not understand exactly what these words meant. Then, with the help of a slick salesman, I learned these words were next to magic for a deer hunter.

This would be a magic potion because buck deer went berserk at the scent of an estrus doe. They would not stop at nothing to find what was producing the scent of love making. The salesman also informed me that this product could possibly endanger me: If I should spill some on myself, the buck might attack me during his rage! Man, was I naive or what!

My ignorance was soon replaced with a little knowledge after a few outings with the magic bottles. That's right, I stated bottles! I was spending just about every dollar I could get on the magic potion. I was going in debt daily while Mother Earth absorbed my allowance for the upcoming weeks.

Today I laugh at the days of my youth and some of the things I have done. There is no question in my mind that I asked for everything I have received. This is very true of some of my past sporting goods bills.

Like many hunters I was merely trying to increase my odds for success without having to think. This is the reason

I advise all hunters to remember that the ultimate piece of their hunting equipment is their *brain*!

Hunters who are consistently thinking never get suckered as I did at one time. Hunters who are always thinking never create hunting accidents! Remember, hunting accidents don't just happen, they were created by the hunter! Remember also that there are various types of attractants and each has its own purpose. This is the reason I feel many hunters create the negative results they do while using deer lures.

HOW ATTRACTING ARE ATTRACTANTS?

Today, after decades of using, experimenting, cussing and praising commercial deer lures, I still have a lot of questions in my mind. Questions such as, why do lures that look better than others do not work as well, or don't work at all? I have used some lures which simply gave no results whatsoever. Today the majority of these lures have fallen to the wayside in the commercial market. Therefore, I assume other hunters had the same experiences and stopped the cash flow.

Many of today's scent companies offer a complete line of attractants. When I make the statement "complete" I am referring to all the attractants needed for luring deer.

Unlike some people who have stated that one type of attractant will do it all, I disagree.

The reason I disagree is what I have been saying all along: Attractants are used to communicate and hopefully lure the deer to the hunter. Therefore, the attractant must represent something natural to the receptive deer. This is the reason hunters should possess a working knowledge of the animal's habits at the time the attractant is being used.

Today's hunters will discover an assortment of attractants on the shelves of sporting goods stores.

WHICH ATTRACTANT IS FOR ME?

For a number of years I have traveled around the country lecturing on scents and their uses. During these seminars I have also enjoyed meeting and talking to thousands of hunters who are quick to point out they feel the same as I do about a certain subject. The subject is scent education. Like most of my acquaintances, I feel very little instruction has been given on how to use attractants. We also agree most of the information is primarily geared to how to use more of it and that's it! Therefore we will now proceed to explore which attractant is best to use and when.

LET THE DEER INFORM YOU

In my opinion, the key to deer hunting is knowing the sexual status of the animals while we are hunting. I base

Scouting is an important element in proper scent usage.

this opinion due to the three basic functions of the deer, which are eating, sleeping, and reproducing the species. One of these functions will motivate the animal to its actions. It is also my opinion that of these three factors, reproduction is the most influential.

Unlike some animals, this sexual influence will be during specific time frames which are commonly referred to as the "Rutting Period."

The rutting periods of the deer are influenced by the number of receptive (estrus) females; the state of their estrogen levels within the given the area. It will be the number of estrus females which will also influence or determine what stage of the rut is in progress.

The various stages of the rut are the key to hunting deer. This is why most hunting seasons are based upon a specific stage or stages of the rut. Please note I stated "most" hunting seasons. Today there are some states which begin hunting long before any form of rutting activity begins.

Today, however, the majority of states focus a specific hunting season around one of the three rutting stages. The three stages I refer to are (A) Pre-Rut, (B) Peak-Rut and (C) Post-Rut.These time frames will vary within the different geographical locations of North America; therefore, it is up to the hunter to know when these time periods are within the given region. This information can also be gathered by scouting the hunting area.

Scouting the hunting area will quickly inform a woodswise hunter as to the state of the animals. Scouting the area and searching for buck scrapes is one of the best ways of knowing it is time to use estrus types scents. This is due to the buck's eagerness to begin the reproduction process of the herd. In most instances hunters will discover these scrapes approximately 10 days to two weeks before the "Peak-Rut" begins.

HUNTING THE PEAK OF THE HUNT

Hunting the peak of the rut is, without question, one of the prime times to be in the woods. For most hunters, this rutting period is when the ol' buck drops his guard. Hunters often experience bucks simply throwing caution to the wind and committing fatal mistakes. These mistakes are made by being more involved with the scent of an estrus doe than in staying alive. Many hunters use estrus doe urine for luring peak rut bucks.

ESTRUS DOE URINE

Unlike some people I do not believe all I read about estrus-type urine. I am not stating that estrus urine lures are not effective; I am simply stating that I do not feel they are as effective during the peak of the rut as during other periods of the rut. The reason I feel this is because, it *is* the *peak of the rut*!

It is this period when the majority of does are in estrus. Trying to interest a buck with an application of commercial estrus doe urine will be difficult. In most cases, the bucks have already located an upcoming respective doe and are in the process of tending her.

In most cases these types of does are in an abundance and only add to the competition. In short, it is hard to fool Mother Nature, but it can be done as we will see in future chapters.

NON-ESTRUS DOE URINE

Hunters poking around sporting goods stores may discover there is another form of doe urine—urine collected

The use of buck urine produced young Douglas Mauldin his first buck.

while the does are not experiencing any stage of the estrus cycle.

This form of attractant is often avoided by hunters because of the non-estrus factor. In my opinion, many hunters avoid a true odds maker. These types of attractants can and will produce positive results when properly used.

BUCK URINE

Hunters often confuse buck lure with buck urine. In my opinion they are two within the same. Buck urine is simply deadly on buck deer during certain phases of the rutting period. Buck urine is exactly what is says it is—buck pee. Therefore, hunters using buck urine should not expect to lure to many does. But if antlers are on the agenda, buck urine is hard to beat in my book.

Attractants are not a "blue pill" for success—they must be used properly and naturally in order for them to have the desired affect. (Photo by Judd Cooney)

MIXTRIXS

Earlier in the book I expressed my feeling towards attractants made from glands.

I simply cannot bring myself to have a lot of faith in these products, but this is only my opinion. I have seen some splendid bucks which have fallen prey to the aroma of the mixtures. I also know some hunters who prefer these lures over all other types. These hunters have the key ingredient which make these lures work so well. They have *confidence*. So, if you have confidence in these products, by all means stick with them. If you are using them and not experiencing much success, why fight it! There are a lot of other good products on the market.

In recent years we have seen synthetic deer attractants become available on the scent market. As with some other types of attractants I feel it is hard to compete with Ma Nature. I have, however, witnessed a few deer respond to some of these blends. In fact, I once had a better than average seven-pointer allow me two shots with a bow at him. The second arrow proved to be lethal while his interest remained on a scent trail of Diz's deer lure.

The seven-pointer has not been the only deer I have harvested while using a synthetic lure. During the past couple of decades I have harvested a dozen or more deer with these lures. It has been from these experiences that I believe the attractants perform better during a certain situation.

The situation I refer to is during periods of heavy hunting pressure. When pressure is heavy, deer may encounter extreme levels of commercial lure scent in the woods.

This is due to the high percentage of hunters all using attractants of the same nature. Therefore, the animals may learn or become educated to these common scents. This is why I try to find out what the majority of hunters are using within a given area. This information can easily be obtained by checking either with the hunters or by checking the shelves of your local sporting goods store.

CHAPTER NINE

PREPARING FOR SCENT STRATEGIES

How many times have you asked yourself, "Why was I not prepared to do that?" If you are like me this is almost a daily occurrence during hunting season. Seems like there is always something I wish I had done before it came time to go hunting.

Past experiences have taught me there are many things hunters can do before the season opens. Things which will increase our odds for success and make hunting more enjoyable.

Things such as having all of our equipment prepared. In my opinion, equipment preparation goes beyond the gun or bow. Equipment preparation includes everything—every item which will be used while I am in the field. This includes many of my scents and lures.

Hunters may find a special kind of cover scent may be needed to properly hunt a location, and this special cover scent may not be available from a scent producer. I have, on occasion, discovered that if I had a certain type of cover scent the hunt would have been better. This is the reason I often formulate some of my own scents.

Formulating special purpose scents has been a hobby of mine for many years. In the beginning chapter I mentioned the old barn which was once my private laboratory.

Inside the walls of that old barn I experimented with

some of the God-awfullest smells one could imagine. Smells were created from my assortment of various animal parts, urines, various plants, you name it, I collected it!

The collection also included some of the favorite foods I knew deer liked. It was with these foods I would blend what I felt would become a sure-fire lure. In most cases I thought more of my special mixtures than the deer did, but there have been a few blends which have paid off for me.

My first payoff came during the early days of the archery season decades ago. I had located an area on the edge of a soybean field the deer were tearing up. The location was in a low valley-like area which bordered a large pasture. The only thing which divided the two fields was a old fence row.

The fence row was like many common to southern rural farms, overgrown with small trees and vegetation. Much of the vegetation was in the form of honeysuckle vines. Honeysuckle vines usually grow to develop thick clusters which cover the ground completely. Often these vines will grow upward along small trees and upon the wire of the fence. In many instances hunters will discover two important factors pertaining to deer and honeysuckles.

The first factor is deer will often browse on the small green leaves of the honeysuckle plant. The second factor is honeysuckles can create excellent ground blinds from which to hunt. In this particular case, the deer would sometimes feed on my blind.

The reason the deer were feeding on my blind was the honeysuckle had grown in such a manner it presented a wall from which I could sit behind. Sit behind and hope the approaching animals would not detect me.

More often than not, the approaching deer would smell me. This was because the predominate wind was usually not in my favor. Many times I thought the wind would

The author learned at an early age to improvise when venison was on the menu.

Bowhunters like David Hale know the importance of being patient if a trophy buck is to be harvested.

simply change direction to irritate me. If I had not seen so many deer feeding in the area, I don't think I would have hunted the place. But I had seen some real dandies and I was determined to arrow one!

The days quickly turned into a week and I had yet to fire a shot. Each and every encounter ended like the one before, in that the deer were spooked. I was beginning to think the deer had become wise to my cover scent. I had used so much skunk scent that even I was becoming used to it. There was no question in my mind, something had to change.

The change took place one morning while I sat behind the wall of honeysuckles. The deer had not moved much that morning and I had a lot of time to think about things. Things such as why the animals had not moved, and why they had. It was while I was trying to unravel these mysteries that I came up with an idea.

My idea was to create my own honeysuckle cover scent. I felt I had everything I needed after filling all my pockets with freshly picked honeysuckle leaves. So with all my pockets bulging, I headed for the laboratory.

In my lab (the barn), I began formulating my secret scent. I first began by washing the honeysuckle leaves in cold water. Then I allowed them to dry while I secretly borrowed my grandma's food chopper.

The food chopper did an excellent job of turning the honeysuckle leaves into a green mush. The mush was than placed in a clean pint fruit jar. I then added a very small amount of glycerine to the mush for my base. This I placed in a dark corner of the lab where the blend would ferment slowly.

The fermentation process took about three days before it received my blessing. The aroma of the blend was really surprising to me. The fragrance of honeysuckle was even better than I had expected it to be. Now I could conjure up one of those sly bucks with ease, I thought.

The following morning found me nestled behind the wall smelling rather pleasant. My confidence was extremely high that morning as I continuously checked my special potion. It was during one of these periodic checking that a buck appeared from nowhere.

The shock I experienced when I turned and looked through the fence was electrifying. Standing less than ten yards from me was a beautiful, wide racked eight-pointer. The buck was nibbling the leaves of soybean and had no suspicion whatsoever.

With the first sight of the buck my nerves began to quiver a little. My shaky fingers slowly found the bow string as I never allowed my eyes to leave the buck. Everything appeared to be in slow motion as I began drawing my bow.

Looking through the opening I had prepared, I had a clear shot into the deer's vital area. This area on the deer is where my sight would remain for long agonizing seconds.

Time seemed to stand still as my bow reached the point of full draw. The buck had lowered its head to gather another leaf when I released the arrow.

The seconds which followed the releasing of my arrow were moments of mental blurriness. Like a lapse in time, I cannot recall the flight of the arrow. Perhaps it was because of the arrow's speed, or perhaps it was my full attention being on the buck's vital area. I still don't know. What I do know is that, for only a brief instant, I remember seeing the arrow fletch disappear inside the deer.

The silence which had been almost deafening was disturbed by the sound of the broadhead striking the buck. The buck reacted instantly to the arrow and bolted forward. With this movement, I lost sight of my trophy. I could hear only the sounds of the escaping buck, and then only momentarily, before total silence.

I tried to react to the buck's disappearance by striking my head through the opening in the fence. This gained nothing more than my cap falling from my head as my nerves exploded.

Within seconds I managed to crawl through the tangle of honeysuckles and enter the soybean field. In the field I realized I had just shot one of the biggest bucks I had ever encountered. Now I was really nervous and it was showing, I had not recovered my cap.

Knowing I had made a good shot I began to relax a little, very little, but a little. I then calmly proceeded to walk to where the deer had been standing. It was then that I spied the scarlet line on the soybean plants.

The appearance of the blood simply sent me into an-

The author advises using limestone pebbles or applying scent in rainy weather.

other world as my eyes followed the trail, which I knew would lead me to what I had dreamed of for many years.

The following quarter of an hour was spent with me counting each tick of my watch. I felt it had been the longest 15 minutes I had ever experienced. It was during this time period that I saw two does cross the far end of the soybean field.

The sight of watching the does shine in the early morning sunlight was very enjoyable. They appeared magnificent as they bounced above the soybeans. These does also alerted me to the presence of my trophy.

The does had almost reached the edge of the field when they suddenly stopped. Like statues they stood for long moments staring away from me, and in such a manner that I could tell they had suddenly become cautious towards something. Their encounter lasted for perhaps a minute or two before they slyly entered the wooded area at the back of the field.

A scent strategy must be based on what the deer are doing—what they're eating, the stage of the rut, how they're moving—at that particular time of season. (Photo by Judd Cooney)

I now understood that the reactions of the deer were due to something that was out of place at the edge of the field. If my suspicions were correct, I would not have to spend time trailing my buck, because the does had found him for me.

In less than 10 minutes my suspicions were proven correct. I allowed myself a little time for some celebration. Then I field dressed the buck and began dragging him home.

The memory of that buck will always burn in my memory. Not only was the buck one of my true trophies, it was one of the first I had fooled by preparing my own cover scent.

Hunters will discover that different scents may be needed in various geographical locations. In many instances, cover scents can be developed from plants native to the exact hunting location.

Hunters will also discover certain preparations can be used to make some attractants even more effective. This is why I feel it is always important for hunters to think ahead of time.

In some regions of the country, hunters will be deer hunting during periods of high precipitation. During periods of rain or snow, most attractants will deplete quickly. Simply, moisture will wash the attractant away. This, in turn, reduces the hunter's odds for successful hunting.

Preparing for wet weather hunting can be done weeks in advance quiet, and quite easily. The preparation can be made with a couple of clean glass jars and some limestone pebbles. I fill the jars approximately one-fourth full with the pebbles. Then I pour my attractant over the pebbles.

In most cases a pint-size jar will need approximately two ounces of urine. Then I seal the jar tightly and allow the pebbles to absorb the urine for about two to three weeks. This will allow the pebbles to contain the scent of the attractant for a long period of time.

The scent used to treat the pebbles can then be transferred to their original containers and the pebbles can be placed in zip-lock bags. Some of the many uses for the pebbles will be described in the following chapters.

CHAPTER TEN

SCENT APPLICATION

The ground was covered with a heavy frost the morning I began exploring scent applications. In my memory I can still see the small rack of the six-point buck. The early morning sunlight was reflecting off the deer as it sniffed the ground where I had previously made my scent trail. The scent trail was supposed to guide the buck to me, but this was the first buck I had ever seen reacting to a scent trail.

Long moments passed as the buck traveled four or five steps towards me. Then, as if something was wrong, the deer reversed his movements and went the other way. Back and forth the deer traveled for approximately five minutes before it decided on its direction. His decision startled me.

I felt that in any moment the buck would turn around and come to me. My hopes were quickly deflated as I watched the buck retrace the steps I had taken coming to my stand and then disappear.

Because of his disappearance, I realized that something was definitely wrong. There was no question in my mind that the deer had smelled my scent trail, but why had he not followed it?

The remainder of that morning's hunt was one of deep disappointment as I did not see any more deer. I returned home and hurriedly accomplished my chores so I could go back to hunting.

The afternoon hunt was conducted in the same manner

as the morning's hunt had been. Like that morning, a deer came from the woods and intercepted my scent trail.

This time it was a small forkhorn. Like the six-pointer, the forky followed the trail in the wrong direction. Now I was becoming very irritated with my scent and the deer.

I had followed the instructions that had come with the attractant to the letter. I had placed the lure on my scent pad at the road and proceeded to my hunting stand. At the stand I had removed the scent pad from my boot and hidden it in some leaves. This was where the deer were supposed to find the end of trail. This was where I was supposed to end the trail for the deer.

That night the occurrences of the day danced in mind. I simply could not understand why the deer had followed my trail the wrong way. Then, as though a bolt of lightning had struck me, it occurred to me: I was leading the deer away from me. How? *Scent molecules*!

Hunters using a foot pad for applying scent should realize something. When we apply the scent to the foot pad we are placing a certain amount of scent molecules on the pad. Hypothetically, let's say we have placed 10,000 molecules on the pad. With each step we take, what are we doing to the number of molecules upon the scent pad is, we are depleting them.

Therefore, with each step we take we are placing a lower number of molecules within that step, and along our trail. Now, we should realize the lowest amount of scent will be at our hunting site, the highest will be where we placed it! We should also realize that if the deer encounters the source of the scent, it has a brain! A brain which will inform it that, by following the weaker amount of scent to the higher amount of scent, it will discover the source of the scent!

The following morning proved to me that my idea was

For whitetail bucks to be harvested with scent it must be applied correctly. Gus Pieralisi, Jr. shows off his skills.

correct. I went to my treestand, placed my scent on the foot pad and walked to where I had been seeing the deer. Then I removed my foot pad and returned to my treestand.

It was not very long after sunrise when I noticed a nice seven-point buck had entered the field and then began walking away from me. Then, as though a neon light had

The author believes many hunters do not use foot pads correctly—it's all too easy to send the wrong message!

attracted the deer's nose to the ground, it began sniffing where I had traveled with the foot pad.

The animal's first reaction was much like that of the others. It moved up and down the trail. Then, with the line of direction determined, the buck began quickly walking towards me. I cannot remember if the buck ever raised its head from the trail. I can, however, remember the buck never raised its head again when it reached the end of trail!

STATIONARY APPLICATORS

Applying scents from stationary applicators can be very productive. Stationary applicators can also save a hunter money. One the best methods for applying scent with a stationary applicator is by using 35mm film canisters. If you recall I reviewed how the canisters can be used for dispensing cover scents. They can also be used for the dispersement of attractants.

When using film canisters for dispensing attractants, I recommend placing them away from the immediate hunting site. I also recommend hanging the canisters about five or six feet above the ground. This I do to allow the animals to walk under the scent so they will not pinpoint the source.

I also recommend placing the canisters approximately 10 or 15 yards from the stand site. This way, if a non-target animal responds, the chances are lower that the animal will detect the hunter.

THE SCENT POLE METHOD

I have stated I learned a lot about scent from trapping furs. A trapper's cash flow depends on how well he/she can lure a critter to the trap. This is also true of deer hunting.

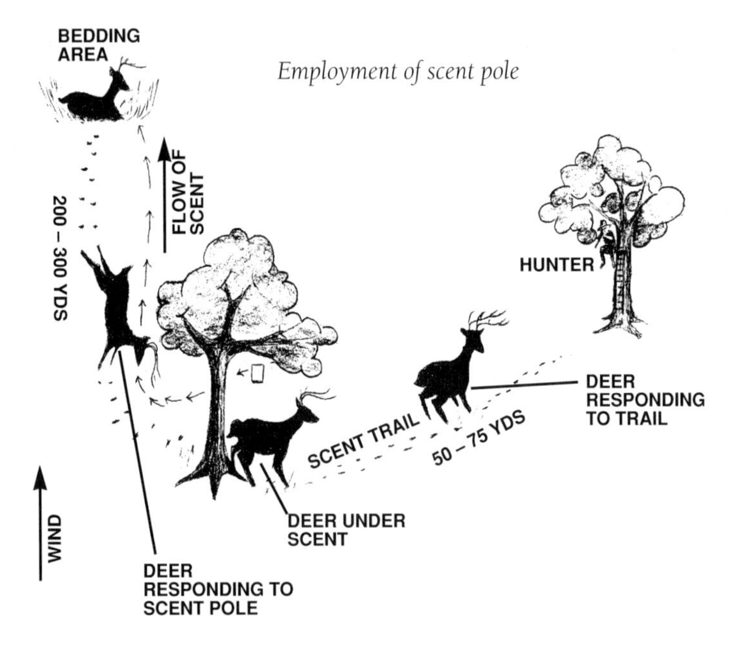

Employment of scent pole

Due to my trapping experiences, I learned various methods of luring deer. One such method was the scent pole method.

This method is nothing more than a long range method for luring deer. This is also a very simple method to incorporate with other scent techniques being used.

To perform the method of scent pole luring, all that is required is the attractant, a clean cloth, and a small tree. The small tree should be as high as the hunter can bend over. When the tree is bent over, attach the cloth to the top of the tree. Place the attractant on the cloth, and then allow the tree to slowly resume its natural height.

When the tree has regained its height, the attractant will obviously be cast at a higher level. This, in turn (and de-

Foot Pad Employment-It's easy to give the deer the opposite message!

pending upon the velocity of the air currents), will carry the scent to a longer range.

In most instances I will try to place the scent pole at least 20 or more yards from my hunting stand. This insures my scent will not be transmitted with the attractant.

I have often used this method while hunting with archery equipment. When doing so, I also incorporate the use of foot pads to lure the deer to within the final distance of my stand. This is achieved by the approaching animals walking under the line of trees at the scent pole.

The scent pole method has proven countless times to be a very effective method for luring deer from their bedding locations. This has been accomplished by placing the scent in such a manner that the air currents will carry the scent into the bedding area. This allows the attractant to appear as a stationary form of scent. In my opinion, this represents another bedding deer.

The only caution I advise is for hunters to locate the hunting site in a manner that their scent will be dispensed by a downwind or crosswind.

GENERAL HOSPITAL IN THE WOODS

I will never forget how excited I was the first time I read about dispensing scent with an I.V. bag. It sounded so good I was sure trophy bucks would be waiting at every scrape. I must admit that there were a few bucks which I think lost their antlers to me because of the I.V. method. There is also one monster buck I was close to getting that the bag saved. How?

Like many hunters I simply hung the bag filled with my attractant in plain view. I never thought the buck would realize the bag did not belong. Perhaps it wasn't the bag, but I knew the buck quit using the scrape when I applied the bag.

When using any form of long term dripping, I recommend placing the dispenser so it is not obvious. In fact, when I do use this form of dispensing, I camouflage the bag. I place camouflage cloth around the bag and add some natural foliage. Once again, I think the deer are not as dumb as some people believe.

THE SUPER-SENDERS

A few years ago I was introduced to the gadget which was going to change the lives of every deer hunter, the high-powered scent dispenser. These little devices would keep the scent warm and natural-smelling and blast it into the next zip code.

The author has lured many big bucks while using the scent pole method of applying lures.

Like many hunters I rose early one freezing morning, got my scent sender and went out to shoot a monster buck. Unfortunately, I did not see a monster buck that morning, and in some ways I was glad of it. I felt this way because of what I learned from the only two deer I did see.

Hours had passed since I had placed my scent sender in

Deer can detect even the slightest noises from long distances—and their response is predictable! Photo by Judd Cooney

a tree approximately 50 yards from my treestand. I was beginning to think every deer in the country had migrated south for the winter. Nothing was moving, not even the birds. I had almost reached my saturation point with being bored when I noticed a doe coming up the trail.

The doe had her nose slightly raised and I felt she had gotten wind of my scent. Watching the doe more carefully I realized she had smelled the lure and was reacting to the device. "This is great!" I thought as I continued watching the deer.

Slowly, but surely, the doe kept approaching the location of the scent sender. It was not until the doe had come within 10 yards that she suddenly halted.

The doe stood motionless for several minutes with her ears in a very alert position. If I had not been watching so closely I might not have noticed the doe look up into the tree where I had placed the scent sender.

The doe's reaction was quick: she turned and bolted away. I observed the deer's reaction as she simply tracked the scent and figured things out. I kept this thought until another doe came by later.

I had watched the doe cross the woods and knew she was upwind of the scent sender. I placed little value upon the deer until she, too, came to a very sudden stop. The deer simply slammed on her brakes approximately 10 yards from the scent sender.

Like the first doe, this doe's ears suddenly became erect for a few minutes before she, too, bolted away. It was from this that I knew the deer must be hearing the sounds of the battery-powered scent sender!

In minutes I had climbed from my stand and was standing directly under the scent sender. I listened closely for any sounds the device was creating, but I could hear none!

Feeling that I had already compromised my hunt by climbing down, I decided to call it a morning and started to retrieve the scent sender. It was not until I had the device within a few feet of me that I could hear the humming of the tiny motor.

Knowing that the hearing capabilities of a deer are much greater than mine I decided the sender would be used for only special occasions. This is why I advise hunters to listen to what they are buying before slapping the cash on the counter!

Today there are numerous ways of dispensing commercial scents. There are a few more methods I use which I will relate in some upcoming chapters of this book.

To sumarize this topic, I simply want to recount that I feel that the key factor to applying our scents is to remain as natural as possible if positive results are to be expected.

CHAPTER ELEVEN

BOWHUNTING THE EARLY DAYS

In most states the first hunters afield will be those using archery equipment. These hunters will often be hunting during periods of warm temperatures which will not only effect the hunter, but the deer as well.

During periods of extreme warm temperatures, hunters will produce more scent due to perspiration.

No matter how hard we try to remain cool and calm, it sometimes seems impossible. This is why being very scent-conscious is important to hunting success.

Being scent-conscious requires the hunter to think beyond just playing the wind direction. It requires the hunter to think about what he/she is doing at all times. We should strive to consider what we are doing and how it will effect us in the future. In short, we should plan every move before we do it!

Planning our movements in advance will simply increase our odds for success. Planning things such as how we will travel to our hunting site is very important during warm temperatures. By planning our travel, we can reduce the amount of scent we carry to the stand. This is the reason I select the quickest and easiest possible path.

By selecting the quickest route to the stand I reduce the time I will be walking. Reducing the time I spend walking reduces exertion. The reduction of exertion simply means we are reducing our perspiration.

The type clothing we wear is another important element in scent production. During warm weather I prefer to wear the lightest clothing possible. Lighter clothing will allow body heat to escape more freely. This, in turn, also helps to control our perspiration.

In recent years I have found wearing the super light camouflage mesh materials to be the ticket. This type of camouflage suit is also inexpensive and durable. I always prepare these suits before hunting season with a natural cover scent, then place the suits in large zip-lock bags for convenience.

Preparing for warm weather hunting is a major factor in hunting success. Preparation is the key to eating venison, in my opinion. This is especially true of the attractants we will be using.

During the early days of archery season the basic attractant I use is *non-estrus* urine. The reason I prefer non-estrus doe urine is because the estrus cycle period is months away. I believe the animals know this and I want everything to resemble a natural state.

In preparing my scent for warm weather hunting, I dilute the urine with a lot of spring water. In fact, I want the aroma of urine to be barely present when I smell it. This is due to respecting the animal's olfactory abilities as I do. This respect came early one morning many years ago.

It was during the mid-1970's and the opening day of the statewide archery season was more than warm, it was *hot*! The early morning temperatures were hanging in the upper 60's. The mercury inched upward with each hour of the morning. By the middle of the afternoon the temperatures where in the 90's. Needless to say, the animals were not moving much in those temperatures.

With today's liberal bag limits, many bowhunters take advantage of every situation. Bill Baird and Mark Atchley with two nice swamp bucks.

Like many bowhunters in the area, I had ventured afield with high hopes and had returned miserable. The source of my misery was from donating my blood to the mosquitoes and, of course, not having seen any deer.

Even with the lousy conditions I still placed my attractants in hopes of luring a deer. I was on the verge of deep frustration when I finally spotted a deer. At first I could not define the sex of the deer until I finally saw its antlers. It was an average sized-spike buck with antlers extending 10 or so inches above its head.

Though some hunters would not have given the buck a second look, I did! I simply wanted a shot at a deer, that simple! I also had no ambitions about trying to eat antlers, I wanted *meat*!

Nearly a quarter of an hour ticked by while I watched the spike browse around. With the direction it was going, I knew it would soon be encountering my scent trail. I had

Scents and hunter confidence are the keys to harvesting trophy bucks.

high hopes the buck would smell my attractant and come charging to me.

I was preparing for the shot when the buck came to the place where I had laid my scent trail. With its nose to the ground, I watched the buck suddenly stop and smell. I had fantasized part of the unfolding episode correctly when I had imagined the deer charging . . . but this one charged for a hasty retreat in the same direction it had came from.

Long minutes passed while I stood in shock at the reactions of the deer. Something was definitely wrong with my scent. I took the bottle from my pocket and removed the top. It was with that first blast that I realized what I had done. I had overpowered the deer. The concentrated urine had simply represented an unnatural situation to the spike.

The rest of that morning was spent thinking about the best method for diluting the bottled urine. I figured that natural spring water would be the best route for the dilution. This proved to be sound thinking on my behalf that very afternoon.

The woods resembled a hot, steamy jungle. The sound of mosquitoes and other insects filled my ears. There were nearly no air currents whatsoever. In short, it was less than ideal conditions for hunting deer.

I was now determined to beat the heat and collect me a deer. I was tired of the heat and ready for standard hunting conditions. I also knew it really didn't matter what I wanted. I was going to hunt, and these were the conditions I would be hunting in.

Perspiration flowed from me by the time I had climbed into my homemade platform stand.

I had taken my time, but the heat and humidity were simply too high, so I sweated. A lot.

I sweated for hours as the sun began sinking in the western skies. This was somewhat of a relief as the setting sun lowered the temperatures. It was also lowering the amount of light inside the woods. Now I would have to depend on my hearing for early detection of an approaching deer.

The early evening sounds of the woods—frogs, owls and a lone coyote howling—were reverberating when I heard a distinctive sound. It was the sound of the dry leaves rustling behind me.

The sounds of the leaves being disturbed sent a flow of excitement charging through me. My excitement grew as the sounds grew louder. I knew this meant that whatever was disturbing the leaves was getting nearer to me.

Slowly I attached my string release to the string of my bow and prepared myself. In moments I could sense whatever was coming toward me would be at the base of my tree. This proved to be very accurate when the sound of a snapping twig sounded beneath me.

Below me in the fading sunlight I instantly detected the ivory form of deer antlers. The whiteness of the buck appeared to almost glow as the deer slowly continued its walk which, within seconds, would allow it to escape.

During the few moments it took the buck to walk from beneath my tree, I had slowly raised my bow. Now with the bow in position I began to draw the bow and hope that the buck would present me with a good shooting angle. It was at the moment I pulled the bow approximately one-half of its draw, that the buck suddenly stopped.

In the fading light I could see the buck's nose was lowered to the ground. This presented the buck to me at a three-quarter going away shot. At 15 yards, this was the type of shot many hunters dream about.

The angle of the buck proved to be just what the doctor had ordered and I released the arrow. The arrow's fluorescent green nock appeared like a shooting star upon its release. A shooting star directed precisely behind the buck's shoulder.

My eyes never strayed from the nock as I watched it disappear in the deer. With the disappearance, the sound of the broadhead striking the buck quickly reached my ears. It was at this same instant I saw the buck lunge upward and forward as it ran away.

Within seconds the sounds of the fleeing buck vanished as I listened to total silence. In the silence I began to shake with excitement. I knew I had performed a good shot and I should begin tracking the buck.

I scurried from my treestand and began walking to where the buck had stood. When I arrived at the spot, I realized why the buck had suddenly stopped there. The buck had stopped because it had smelled some of my diluted doe urine. The location had been one of several where I had simply applied a dozen or so drops of the attractant on the ground.

Hunters can create many of their own lures from natural ingredients such as acorns.

I had made this application in numerous areas before I had entered my treestand. I felt this application would capture a deer's attention. I also thought that if the animal's attention was focused on the ground, it would not detect my position. Now I knew my theory was correct.

The tracking of the buck was quite easy as the arrow had made complete penetration. The buck had traveled less than 100 yards after both lungs had received the broadhead.

Today that splendid nine-pointer stands as one of my all-time big bucks. That buck also stands as the first one of many deer which I have shot while they were smelling a diluted attractant during warm weather.

Experimenting with various scents has presented me with a lot of deer. Through the years I think I have tried just about everything imaginable for attracting deer. Everything from animal urines to various forms of food.

FOOD ATTRACTANTS

Food attractants can be a very productive way to attract deer. This is especially true during periods of warm temperatures. As I have stated, deer can smell better during periods of warmth and humidity. Unlike urine attractants, food attractants are not as strong smelling. This is the reason I often use a food attractant for early season bowhunting.

Today hunters can find an assortment of food attractants on the commercial market. Most of these products are very realistic in their appearance. If used properly, food scents can be an effective tool for deer hunting. I learned this years ago while experimenting with an idea I had conceived.

The idea came to me one day while I was sitting in class. Finding the subject at hand somewhat boring, I began plotting for that afternoon's hunt. I had not seen very many deer and figured it was due to the extremely warm temperatures we were experiencing.

My idea came about when the teacher began telling us the process for producing peanut butter. I listened closely to my instructor and decided I would develop the ultimate deer lure, acorn butter!

That afternoon I gathered a large quantity of freshly fallen acorns before entering my tree-stand. Like the evenings before, I went home without seeing a deer. This did not bother me much as I was more interested in developing my secret lure.

That night I crushed the acorns until they were finely ground. Due to the low moisture content of the acorns, they were almost nothing more than a powder when I finished. To this powder, I added a very small amount of dark Karo syrup.

The blending of the syrup to the acorns soon became a very thick glob. The glob was then placed in a clean fruit

Few things are more difficult than early season, hot-weather bowhunting. The deer will move—but it takes careful hunting and creativity to make it happen. (Photo by Judd Cooney)

jar and the jar was placed on the shelf. The remainder of the night was spent with me dreaming about the monster bucks which would come running to my special blend.

Like always, the school bus ride the next afternoon was the longest I could remember. The dust from the bus had barely settled before I had the jar and was headed to my treestand. When I arrived at my hunting site, I took a stick and began smearing the acorn paste on the leaves beneath several of the oak trees surrounding my tree-stand.

I did not see a deer for the remainder of the evening. In fact, I saw nothing for the next three afternoons. It was not until the fourth afternoon that I experienced a shocker.

This experience occurred soon after I entered my tree-stand. I had hardly settled down when I heard something rustling the leaves behind me. Turning to see the source of the disturbance, I saw a large doe. The shocker was that the doe was chewing a mouth full of dried leaves.

Never before had I witnessed a deer eating dried leaves. Then I remembered I had placed some of my special blend where the doe was standing. The thought of the doe eating my lure made me very excited.

The excitement only increased as she lowered her head to search among the leaves. This movement would prove to be the deer's fatal mistake.

Since that afternoon I have used my blend to distract other deer. Never again did I see any other deer eat the leaves as that one doe had. I have, however, seen numerous deer focus their undivided attention on the blend before I released the arrow!

CHAPTER TWELVE

THE BETTER DAYS OF BOWHUNTING

The air felt crisp and clean as I marveled at the brilliant colors of autumn. The golden rays of a new sun were reflecting through the tree tops. It was one of those mornings that made you feel glad you were alive and hunting.

Less than an hour had passed since I had watched a small six-point buck walk by. The deer had spent many minutes investigating the numerous buck rubs in the area. Most of the rubs were nearly a week old. Two of the rubs were freshly made by the little six-pointer.

I had enjoyed watching the small buck create the two rubs, which were located less than 10 yards from my treestand. I knew, as I had watched the deer creating the rubs, that I would use them for my own personal gain.

I enjoyed watching the six-pointer and I never thought about shooting him. Actually, I silently thanked him for the rubs as he left. I felt the deer had done me favor and I was waiting for a bigger buck.

Many hunters prefer to hunt animals which are considered to be "trophy class" or "record book" deer. This is great and on occasions I, too, practice trophy hunting. But, I feel that any whitetail deer taken with a bow and arrow is a *trophy*. In fact, I think any adult whitetail deer taken in fair chase is a *trophy*. I do not believe anyone has the right to judge someone else's game, that simple!

I also feel that hunting rub lines is one of the best methods there is for taking bucks. This is especially true during the latter portion of the archery season. It is usually during the latter days of the season that hunters will encounter numerous rubs. This is due to the bucks dissolving their summertime bachelor groups. It is during this dissolvement of the rankings that a pecking order is established among the deer. During this time frame the bucks will be setting the standards for the upcoming rut.

How important is this to the bowhunter? Plenty, in my opinion, and this is why.

During this period mature bucks are highly vulnerable to a crafty hunter. During this two to three-week period many of the bucks are crazy with rage towards any challenging buck who intrudes upon their territory.

Territories are marked primarily by the buck's rubs, and it's from these markings that my favorite bowhunting method comes forth. The method I refer to is the *mock-rub* method. The mock rub method for attracting deer is basically just what it says it is, making mock, or fake, rubs. Hunters can easily imitate these rubs by using a discarded deer antler.

With the aid of the deer antler, the hunter simply rubs the bark off a small tree to create the rub. This becomes a visual attractant to any deer that may be passing by. The influence of the visible appearance of a buck rub is something hunters should take into consideration. The better the a deer can see the rub, the greater the odds for a response from the deer. Therefore, it is imperative to place the mock rub in a location that can be easily seen from a distance.

I also recommend rubbing the entire tree so all sides are visible. In most instances I will use a tree which is approximately three inches in diameter. Three-inch diameter trees

The author feels buck rubs are quick indicators to learning the presence of a trophy buck.

Even top hunters like Harold Knight says, "There's no place like home for hunting big bucks."

seem to be the most preferred by deer during these time periods. I have also noticed that evergreen trees or softwood trees receive the most rubbing activity. This is the reason I select these types of trees when preparing a mock rub. I want everything to be as natural in appearance as possible. Remember, we are trying to imitate what deer are doing! The more realistic we present the mock rub, the greater the odds that the rub will present a deer to the hunter.

DOING IT QUICKLY

Unlike the old days when I could hunt when I wanted, where I wanted, and what I wanted, much of today's hunting revolves around a schedule. There are times when I have a week to hunt an area and there are times when I only have a few days. One day I might be hunting the vast lands of southern Texas, and the next, wading snow in northern Michigan.

I am not saying that hunting different places is not fun, but there is no place like *home*! I love to hunt the same lands I grew up on. I simply enjoy harvesting the animals I feel I have helped raise. Like a rancher I want to reap some of the fruits of my labors. This became a problem a few years ago during the archery season.

Like many in the hunting business I try to make every day count during the fall of the year. This particular year I discovered I had over-booked myself. In fact, I had only allowed myself four days to hunt on my home ground.

Returning from a trip, I quickly gathered my equipment and headed to my grandma's farm. The fact that it was raining cats and dogs had little influence on my frame of mind. I was going to locate a spot for my portable treestand and pray the weather man was correct. If he was right, the rain would end by midday and I could hunt.

Knowing exactly where I wanted to place the treestand, I proceeded through the downpour. In the pelting rain I placed the stand where I had collected several splendid bucks over the years. While I was returning to my home, I noticed there were only a few rubs within my hunting area. This was somewhat disturbing to me as I had expected to see more signs of bucks.

My thoughts were on the low amount of rubs I had seen while I ate lunch. With my belly full and the weather man having been correct after all, I headed for my treestand.

The woods smelled fresh and clean after the rain had ended. It was simply refreshing to be in my favorite woods and enjoying nature. I also knew there was a good chance it was going to become even more enjoyable. I felt my odds for collecting a nice buck were rapidly improving with the clearing skies. With the improvement in the weather I began applying my hunting strategy.

The strategy revolved around the changing weather conditions and mock rubs. I felt that during the period of the heavy rains and high winds, the animals had not fed very much. I also thought that the deer would know an abundance of acorns had been shaken from the trees due to the high winds. This, I felt, would produce a high number of animals in the vicinity of the oak trees.

To complete my strategy, I prepared six mock rubs in easy range of my treestand. This I hoped would attract any wandering buck within range of my arrow.

Soon I was high in the treestand full of excitement during the final hours of daylight. In the glowing sun rays, I detected an abnormal reflection to my right side. This reflection quickly received my full attention as I knew what

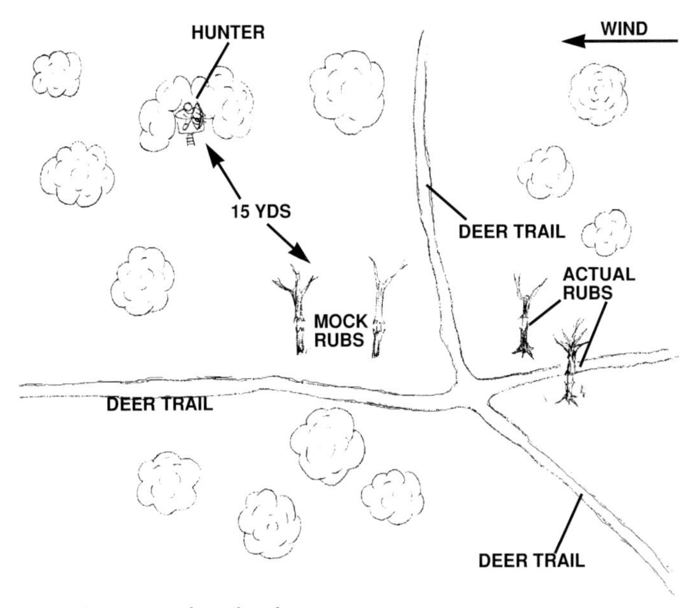

Placement of mock rubs.

had produced it. Like countless times before, the sunlight had betrayed the buck's antlers.

The sudden surge of excitement was quickly battled as I knew I had to remain calm. I could easily see the buck walking behind me from my right to my left side. Watching the deer from the corner of my eye I estimated the range to be close to 50 yards.

I knew something would have to transpire quickly if I was to get a shot at this buck. Tension was now mounting within me as I began watching over my left shoulder. I was becoming tense knowing that the buck was traveling a straight path. I was becoming even more tense knowing the buck was also moving into a downwind position.

I felt the buck would spook any second as I watched it

come to a sudden halt. Like a statue, the deer stood motionless staring in my direction. The buck's actions were becoming more alarming as it suddenly lifted its head and sniffed the air currents.

The buck's action made my heart sink. I felt the deer had detected me because its ears had become full erect. I could see the buck was about to escape when I heard the snapping of a twig beneath me.

At once, I shifted my sight from the buck towards the direction of the noise, and quickly reversed my priorities when I saw the wide rack of another buck beneath me. Due to the rain-soaked leaves, a deer had slipped up on me.

The wide rack of the nine-pointer was clearly visible as I looked straight down at the deer. Knowing this was a very poor angle to attempt a shot, I stood motionless.

Now with my heart beating rapidly, I watched the buck begin to move away from my position. Because of the buck's movements, I quickly scanned over my shoulder for the other buck.

In seconds I had relocated the buck's position and could see the deer was slowly approaching within range of my bow. I could also tell that the first buck was not as large as the one beneath me. No time was wasted in deciding which of the two bucks I wanted, and very little time was wasted in preparing for the shot.

Raising my bow, I noticed the nine-pointer was heading towards one of my mock rubs, and I felt a sudden sense of pride that only a hunter or trapper can feel in knowing they have outwitted their game.

Slowly the buck began smelling the mock rub as I began drawing the bow. Drawing and releasing the arrow was like a normal reflex for me, and sent the arrow disappearing behind the shoulder of the buck.

The reaction of the buck to the arrow was completed in seconds as it bounded forward. The forward motion of the buck allowed me to see my arrow sticking in the ground.

The sight of the arrow in the ground brought forth momentary panic as I listened to the escaping buck. Only the sounds of fallen limbs being broken could be heard before all was silent again.

The brief moment of silence was quickly shattered as the other buck began its hasty retreat. Like the nine-pointer, I followed the deer's path with my ears before all was quiet again.

Time slowly ticked away as I allowed one-half hour to expire before I began tracking my buck. The tracking of the deer was quite easy. The arrow had penetrated the right lung and the heart as it exited the buck.

Within minutes I had removed all of the waxy secretion from both of the buck's pre-orbital areas. With this safely stored in a clean container, I began field dressing the buck.

Soon the task of dragging the deer from the woods to the pick-up truck was completed. Now I could enjoy a cold drink before returning the my stand site. I was going by to the site not to hunt, but to prepare for the next morning's hunt.

The preparation was completed in a matter of minutes. I quickly placed a very small amount of the pre-obitial wax on each of the mock scrapes. Then I sprayed an ample amount of diluted buck urine approximately four feet from the rubbed tree. These applications presented the scent of a buck within the mock rubs. With a little luck there would be numerous other rubs when I returned the next morning. In a matter of speaking, I was simply trying to inform another buck that I had invaded his territory.

IT'S NICE TO BE A BUCK

The following morning found me resting high in my

portable treestand. I had the entered the stand nearly a half hour before dawn. Now the sunlight was beginning to set the stage inside the woods. With the increasing light I could now see my mock rubs—as well as seven actual rubs which had been made within yards of the mocks! My excitement mounted with the sighting of the new rubs. There was little question in my mind that I would see deer that morning.

Less than an hour had passed when I spotted the first deer of the morning. The band of four does slowly entered the area beneath my treestand. Long minutes slowly ticked by as the does browsed among the freshly fallen acorns.

I was enjoying watching the deer as I had no desire to shoot one of them. I felt that if I would be patient, my target for that morning would be carrying antlers.

Nearly a quarter of an hour passed as I watched the does move about the area. They browsed slowly, twitching their tails in a relaxed manner. Then, as if some sort of silent alarm was sounded, all of the does became alerted. Their alertness was signaled with the sudden raising of their heads. Their heads snapped upward as their ears became fully extended. Like statues, the deer stood motionless staring into the direction behind me.

The alertness of the does was now receiving my full attention. I knew the animals were seeing something, which I felt was not a threat to them as they were not frightened, only observant.

Time seemed endless before the does became relaxed again and resumed their browsing. It was because of the actions of the does that I began slowly turning my head to view behind me.

I had no sooner began turning my head when the sight of buck's antlers gained my attention. In the brilliant sunlight the buck appeared to be marvelous. Like some sort of

Experienced hunters like David Hale learn quickly which buck is the biggest when confronted with numerous deer.

Like commercial scents, the author feels commercial grunt calls are valuable tools, also.

monarch, the buck boldly walked towards the browsing does. His actions were that of grace and pride. He acted as though he was without any fears, the invincible king of the woods.

I watched the buck slowly walk in his prideful fashion to within mere feet of my tree. Like the buck from the previous afternoon, he sniffed the ground carefully. He displayed a state of caution as he moved towards one of the new scrapes.

This movement enabled me to assess his rack. It was also with this movement I knew the buck had entered the danger zone: the zone which permitted me the perfect shooting angle, and which would provide me with another lesson in hunting whitetail deer.

In only a matter of seconds I knew I wanted this buck. His rack supported eight highly polished tines. The buck was, without question, a trophy in my book. I began to raise my bow very slowly. I knew that within seconds my moment of truth would be upon me.

It was in the exact moment I began drawing the bow that the buck approached the actual rub. With the buck's nose less than an inch from the tree I learned another a lesson in whitetail behavior.

The lesson was given to me in the form of sound—the sound of a buck grunting. The moment the buck sniffed the rub, I heard the double low-volume guttural "g-r-r-r-rt, g-r-r-rt." Though I have heard deer produce these sounds before, I had never witnessed this at a rub.

The sounds produced by the buck distracted my attention from my bow momentarily. Now I was more interested in learning from the buck than shooting it. Something told me to be patient and I would be rewarded. This quickly proved to be sound judgement on my behalf.

Within seconds of hearing the buck's grunts, I heard a

The scraping behavior of rutting bucks—and the advantage of hunting a scrape—gives hunters able to hunt the rut a tremendous advantage. (Photo by Judd Cooney)

single grunt come from beyond the buck. Now I was becoming very excited as I waited. I knew there was another buck in front of the eight-pointer. Perhaps the other buck was larger?

Tension mounted within me as I watched the eight-pointer stand at attention staring straight ahead. I could sense the deer was going to move as I could see its muscles tightening. Now I was faced with the dilemma of not knowing what was beyond.

My mind was racing as I thought to myself, "If I wait I may miss my opportunity . . . but if I wait, I may encounter an even larger buck!" My thoughts were going wild as I began drawing the bow.

Experience has taught me that a deer in hand is worth more than a herd in the bush. I released the arrow which ultimately presented the beautiful buck which adorns my office wall!

CHAPTER THIRTEEN

HANDLING WHITETAIL TURN-OFFS

Hunting has undergone many changes since the simple days of my youth. Some changes have been good, while others have not been so good. Some positive aspects have been in the equipment that has been developed during the past couple of decades.

Today hunters have some of the best equipment at their disposal—equipment which can be very effective when used by a knowledgeable hunter. Equipment such as game calls, scents, and decoys have been developed to increase the hunter's odds for success.

These products are simply instruments which enable a knowledgeable hunter to communicate with the game. In most cases these products can be effective when the proper conditions exist.

Hunters must understand these conditions if a sound strategy is to be developed. Conditions such as food, weather, hunting pressure, rut phase, moon phase, etc., will influence the habits of the animals. Experienced hunters know how much of the animal's habits are determined by the habitat. Therefore, it is during a period when these factors change that hunters encounter what I term a "turn-off."

A turn-off is a period when the animals are adjusting to new factors or conditions within their habitat. An excellent example of this is when active scrapes suddenly become inactive. The scrapes appear to be nothing more than dirty places on the forest floor.

This phenomenon often will upset hunters who have worked hard to locate a buck's scrapes. I know I have become upset numerous times during what I term as the peak of the rut turn-off.

THE PEAK-RUT TURN-OFF

The peak of the rut turn-off is the sudden period when deer seem to simply disappear. This turn-off is readily identified when the majority of scrapes and primary trails appear to have been abandoned. This occurrence may lasts for only a few days or extend into weeks, depending on various factors.

Perhaps the factor which influences this the most is the number of active estrus does within the area. The more receptive the does in the area are, the less the bucks will investigate their scrapes. This is due to the scrape being created primarily for the buck to seek out a receptive doe. Once the buck finds a receptive doe he will stay with her until mating is finished. When the mating process is complete he will then search for another doe. It is at this time the buck might check his scrapes for a clue in seeking another doe. This is the reason I feel some hunters miss a golden opportunity by not always thoroughly checking the abandoned scrapes frequently.

DEALING WITH VAMPIRE BUCKS

I am often asked what I think makes deer seem to disappear from an area. In response, I simply inform these peo-

The author believes many miss out on trophy bucks when they disregard scrape hunting.

ple the deer don't disappear, they just adjust to existing conditions. One condition that I believe can change a hunting area more quickly than any other is hunting pressure!

Hunting pressure is simply the amount of human activity within a given area. Deer know they must avoid humans and human situations in order to survive. Therefore, the animals seek areas which will provide protection against human intrusion. Extreme hunting pressure may result in the deer becoming nocturnal, or what is considered nocturnal. Personally, I do not believe deer become totally nocturnal. I will agree the amount of night time activity will increase, but not to the degree that some hunters are led to believe. My opinion is based on hunting these so-called vampire bucks for many years.

Today, after years of experience in hunting nocturnal deer, I have not discovered any foolproof technique for hunting them. In fact, I don't know of anything which will provide even odds for hunter success, despite what some gurus say.

I feel it is almost impossible to predict what an animal will do during stressful conditions—especially when heavily pressured. Veteran hunters know, however, that even with the odds running heavily against them there is still a chance for a shot at the big one. This slim chance keeps us persistent hunters in the field trying to unravel some of the mysteries of hunting nocturnal deer.

The value of being persistent paid off a few years ago when I was hunting some inactive scrapes. For nearly a week the numerous scrapes surrounding my stand site appeared totally abandoned by their creator. With the exception of a small four-pointer I had not seen another buck, but I knew a heavy-antlered eight-pointer had been there during archery season and I was in hot pursuit.

Deciding to experiment, I placed some Ambush-brand estrus doe urine in one of the scrapes before calling it a day. Now I could only wait until morning to see if anything responded to my lure. I simply had a feeling that, if the big eight-pointer smelled the lure, his track would be in the scrape.

The next morning I placed a Feather-Flex deer decoy approximately 10 yards from the treated scrape. Then, with the aid of my flashlight, I inspected the scrape. The inspection brought a sudden warmth to me. I beamed with excitement as I found numerous fresh tracks within the scrape. In the pale light of the flashlight the indications were that two deer had visited the scrape during the night.

The remainder of that day was spent viewing only the dummy deer and a couple of passing hunters who were on their usual time schedules going to and from their stands. I felt the residents of the woods knew these time schedules

Being persistent has helped today's top deer hunters get where they are!

felt the residents of the woods knew these time schedules only too well.

With darkness closing in on another day, I retrieved the decoy and treated the scrapes again. This time I incorporated the estrus urine with a hearty dose of buck urine. My reason for the buck urine was to leave the impression that an invading buck had intercepted the doe. If my theory worked, the scrape's creator would become more protective toward his territory. If this theory proved correct, the buck could be in serious trouble tomorrow!

SHOW TIME AT LAST!

The first four hours of the following day were spent gazing at the decoy. The midmorning ritual of hunters returning to their camps was signalled by the barking of gray squirrels. The noisy rodents continued their calling until the hunters had vacated the area.

The woods were finally silent again as the brilliant sun began defrosting my chilled body. The relaxation was brief as the sound of rustling leaves in the distance cut the silence.

For long seconds I listened to the rhythm of the noise before deciding to move. I almost knew the disturbance was being created by an animal as the squirrels had remained silent and unalarmed.

Realizing the sounds were behind me, I slowly began checking over my right shoulder. Now my ears would guide me through the underbrush to locate the polished antlers of the buck.

Excitement raced through me as the buck materialized from the undergrowth. For several seconds the buck stood still and faced the direction of the scrape. Like a statue, the buck stood frozen in its tracks before slowly moving forward.

Ever so slowly did the buck walk toward the scrape with its ears erect. Realizing the deer was now looking at the decoy, I began slowly turning around in my treestand. My actions were quickly halted as the buck suddenly stopped and began raising its tail. For a brief second, I feared the buck was about to attempt escape.

My reaction to the deer's sudden action was quick as I raised my rifle. With the rifle in position I placed the crosshairs of the Simmons scope on the buck's shoulder. I viewed the buck for only a few seconds before squeezing the rifle's trigger.

With the eruption of the rifle the deer dropped to the ground. There was no question in my mind but that the shot had been a fatal one. The buck never moved after the echo of the shot and lay as testimony to successful scrape hunting. Hunting scrapes is the best way I know to even the odds on nocturnal bucks.

LET IT BLOW

Another whitetail turn-off which sometimes baffles hunters is changing weather conditions. Over the years, countless theories have been expressed regarding this turn-off. Personally, I find only one weather condition which is a true turn-off: wind.

Wind affects hunting in a number of ways. It distorts sound, which is negative to both the hunter and the deer.

Wind also disturbs the ever-important air currents, which makes scent control difficult for the hunter. The movement created by wind simply makes deer very nervous and difficult to hunt. Basically, wind is a real pain and

The sight of a big buck is what keeps hunters returning to the woods year after year!

decreases the odds for success substantially. But even with reduced odds, there is always a chance for success for the willing hunter.

During windy conditions I prefer hunting areas which provide food and visibility for the deer. This is due to the animal's sight becoming its primary defense in windy weather. Stand sites located near heavy cover should better the odds. In a nutshell wind, like all other elements in nature, is a natural phenomenon to the deer. They still must eat and sleep even during periods of high winds. Seasoned hunters know this and adjust to the given circumstances. Hunters using scents know they, too, must adjust their scent applications for hunting windy periods.

When hunting in windy conditions I prefer to use undiluted lures. In fact, I save many of my strongest-smelling lures for hunting during these conditions. The reason for this is wind will distort the dissipation of scent molecules. The distortion will be in the form of scattering the scent molecules within the strong air currents. Because of this the animal will not receive as high a number of molecules as it would in favorable conditions.

My preference for applying scents during periods of high winds is using 35mm film canisters. I can hide the little canisters at ground level. The purpose for this is to lure any responding deer into range as the scent will be greater at the scent source.

LET IT RAIN

I do not believe that rain is a real turn-off for deer. It tends to be a real turn-off for hunters, but excepting a genuine torrential downpour (which is generally of too short a duration to be a real turn-off) deer aren't bothered much by precipitation. After all, they live in it year-'round and still must go about their business.

The problem with rain is that it quickly dilutes and dissipates scent. That's when I break out my scent-soaked pebbles. These seem to hold the fragrance better during rainy weather than most other mediums. However, rain will limit the range of virtually all applications of scent; just like visibility is restricted in rainy woods, uses of scent become more up close and personal. The thought of close encounters brings me to the next topic . . . .

PUT IT WHERE IT COUNTS

Experienced hunters know the value of key stand positioning and the factors which influence a stand position. In reality, the stand position is nothing more than the hunter's ambush site. If the stand site does not afford concealment, other negative factors will be introduced. One such factor is movement.

Hunter movement has saved many a whitetail from the deep freeze. Wise hunters should note that once a deer hears a call the caller's position is pinpointed by the animal. Any movement by the hunter can be easily detected by the sharp eyes of a whitetail. The use of quality camouflage always helps in decreasing the odds of the hunter being detected. The odds of a whitetail not detecting hunter movement are slim unless the animal is distracted. Distracting a responding deer has been made easy with the modern deer decoy!

In recent years I have had little trouble harvesting a deer that was approaching my Feather Flex deer decoy. In most instances the fake deer captured the target animal's attention. But I must admit I have altered the decoy in some

The author feels that treestand positioning is as important to hunter success as using scent properly—and the two are tied together.

Apparent whitetail shut-down is one of the most frustrating events for a deer hunter. Changing conditions, whether weather or pressure, is usually the culprit. (Photo by Judd Cooney)

ways. One such way is I apply a healthy dose of non-diluted doe urine. This, I feel, masks any odor the decoy could have on it.

Another alteration is I pin a three-inch strip of white tissue paper to each ear of the decoy. This will create a little movement if there is any air flowing. I feel this slight movement is very important to decoying deer.

Having decoyed numerous deer, I firmly believe the decoy must also be placed in a natural manner; that is, the decoys should face toward downwind. This is important as a resting deer will do this to allow its eyes and ears to protect it where its nose cannot. The nose of a resting deer will detect scent approaching from behind as the air currents drift over the animal's body. Hunters should take note of such things if the stand site is to reach its full potential.

Having talked with hunters all across the country, I have learned how many blame scents for their failures while hunting. Of the many reasons failures may occur, one reason is very apparent among most hunters. It isn't scent or its use or misuse—it's improper stand location.

Numerous times I have listened to hunters tell the story of sitting by scrapes for days, only to look over their shoulder and see the buck escaping. In the opinion of most people I talk with, it was the commercial scents' fault the deer got away. It was the lure which they had placed in the scrape which sent the deer a-running.

In some instances this may have been the problem, as we have reviewed throughout this book. *But* if the hunter saw the buck over his/her shoulder, what should that tell you? The first thing is that the chances were good the deer smelled or saw the hunter if it was approaching from the hunter's back. Second, it should be pretty apparent the treestand was not in proper position.

From my personal experience I have have discovered most hunters place their stands too close to the scrapes. In fact, I have learned most hunters place their stands within 20 yards of an active scrape.

Now, let's think about what is really happening when a buck investigates a scrape. The main reason the ol' boy wishes to check his scrape is to see if a receptive doe is nearby. This the buck can easily do by swinging downwind of the scrape and sniffing the air currents. In most cases I have found that most bucks will approach a scrape from at least 50 or more yards downwind. With this information I hope one would see the importance of keeping distance from the scrape.

In most instances I will place my treestand as far away from the scrape as I can. I like them to be within easy shooting range of the scrape and nothing more!

Hunting late-season scrapes can be the very ticket to placing the buck of a lifetime on the wall. Hunting late season scrapes will also prove to any hunter his devotion and dedication to being a true deer hunter!

CHAPTER FOURTEEN

HUNTING THE HARD TO GET

Spending most of my life in Northwestern Tennessee has been very rewarding for me. Within this region I have access to almost any type of hunting terrain one could want. From rolling hills to farmlands to river bottom swamps, my native land has it all within minutes of my home. It has been this assortment of hunting terrain which I feel has enabled me to become a good hunter. If I had not experienced hunting deer under various conditions, I do not think I would be where I am today. In fact, I believe of all the various types of whitetail habitats, hunting swamplands may be the most difficult.

I began hunting swampland for waterfowl and small game animals at a very early age. The whitetail deer population was very low during this time period, so other critters were on my agenda. I learned many lessons while hunting these animals which would help me in the future. I also learned the value of using various scents while hunting swampland.

The way I began hunting swampland started one cold December morning. I was trying desperately to avoid getting a wet hip boot while wading to my duck blind. The task of staying dry was a lost cause due to a sink hole. The sudden shock of the cold water filling my boot made me freeze in my tracks.

In minutes I had located a muskrat house and was emptying my boot of the cold contents. It was while I was standing on that muskrat house that my life changed.

While I stood there, the eerie sounds played a trick on my young imagination. For a brief moment I thought a herd of cattle was charging me through the water. Then I saw the culprits in the dim light as they came closer.

The first positive identification was the enormous rack of the leader. In seconds the massive rack of the buck was less than 10 feet away from me as it splashed through the water. Behind the buck was another smaller buck and four does. The deer were quickly engulfed in the nearby cane thicket. Long moments passed while I stood in disbelief and listened to the deer fade away.

The remainder of that morning was spent trying to duck hunt with deer on my mind. The sight of the big buck had me ready to trade the shotgun for my deer rifle. I was ready to begin hunting swampland deer.

DEVELOPING A SWAMPLAND STRATEGY

Unlike the woodlands I was used to hunting, I knew swamp hunting would be very different. I knew I would have to use different equipment for hunting the swamps. The use of waders would have to be part of my equipment, as well as a small boat. I also knew that, due to the terrain, the habits of the deer would be different as well. Therefore, I would have to scout the areas extensively for the few clues I would receive. I also felt that the more hunting pressure placed on the surrounding areas would only improve the swamp. I decided to use the swamp for the latter days of the hunting season.

My hunting strategy began with intensive scouting of the vast swamp. I scouted the edges of the swamp on an almost weekly basis. I would scout for all the obvious signs

Swamplands can produce monster bucks, says the author—demonstrated by Douglas and Draper Mauldin—Double Trouble!

of the whitetail deer such as trails, rubs etc. I would then record all my findings on a topographical map I had obtained of the area. For weeks I spent hours walking the outer parameters of the swamp searching for clues. Most of my scouting proved to be only good exercise while slashing through the mud and water. In short, I had found only a few trails and a couple of rubs for all my effort.

I was beginning to think perhaps swamplands were not the way to go for late season hunting. I was considering staying with the farmland deer, even though the hunting pressure was taking its toll. In short, I felt I was wasting my time. And then it happened!

THE BEAVER DAM BUCK

Having spent a lot of time in discovering a way to hunt the swampland, I was becoming weary. In fact, I was beginning to become very disgusted with deer as a whole. It was due to this disappointment I decided to go try my hand at some duck hunting.

Like that first morning I recounted, I managed to get my feet wet before I entered the duck blind. This time I selected an old beaver dam to empty the cold contents of my boots.

During the process of draining my foot gear, I happened to look down on the beaver dam. Lo and behold, the sight I saw was more shocking than the cold water. I was looking at numerous deer tracks which were on top of the beaver dam! I simply could not believe my eyes.

Needless to say, the ducks received very little punishment that morning. I quickly began following the well worn deer trail across the beaver dam. From the beaver dam I could barely see the deer tracks in the shallow water.

Most of that morning I carefully traced the path of the deer across the swamp. It was not until noon did I realize what the animals were doing. They were using the tiny islands or high places for their bedding grounds. I then also realized that the deer were doing most of their traveling at night. They were feeding among the farmlands after sundown and returning in the early morning hours.

THE PLAN UNFOLDS

The next few days found me scouting other areas of the swamp. I knew there had to be more deer there. I also knew there were some key factors to hunting these deer—factors which I would have to learn if I was to become successful.

From the extensive scouting, I located other areas where deer had traveled in and out of the swamp. These areas

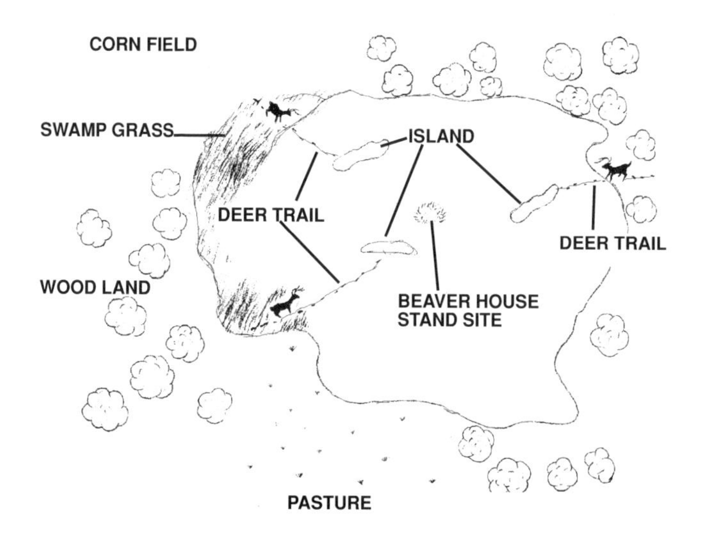

were recorded on my map with the other information. On the map I listed the primary food sources bordering the swamp.

I also made special notes of the shallowest water routes leading from the food sources to the island. I also recorded the locations of the area's landmarks, such as beaver dams and cane groves. The landmarks also consisted of large cypress trees and small patches of water oaks. The large cypress tress would aid me in navigating the swamp. The water oaks informed me of possible food sources.

Now I began making plans to do some serious swamp hunting. I enlisted the help of a close friend to help me transport a small boat to the swamp. With the boat I would now be able to explore the areas where I thought held a high concentration of deer. The first area on my agenda to investigate was where I felt the big buck I had seen resided.

With the aid of the topo map and the boat, I located a small island. The island was located approximately 200 yards from where I had spotted the monarch.

Paddling to the island was an exciting experience within itself. The excitement was generated when a small herd of deer suddenly bolted from the island into the water.

With the sounds of the escaping deer, I knew I had uncovered one of the mysteries of hunting swamp deer. I also knew I had gotten a brief glimpse of a large rack before it, too, disappeared beyond the island.

Upon reaching the island, I realized the markings on the trees were not what I had first thought. Like numerous other areas in the swamp, the lower portions of the trees were void of bark. This is common in swampland due to beaver activity, but what I was looking at here was the result of large antlers scraping the trees.

There was no question in my mind that some pretty impressive bucks had been on the island. The only question which played upon my mind was how I was going to hunt these bucks.

I felt I would have to wait in ambush and hope a buck would wander onto the island. I also knew I was playing a game of chance as there were many of these islands scattered around this vast swamp. Finally, I decided to employ the use of scents.

MOCKING FOR SUCCESS

The sun was hours away from the horizon when I began paddling the small boat across the swamp. With the aid of my landmarks and my compass, I arrived at the first island. Once on the island, I quickly located a scrape which had been recently made by a buck. Within the scrape I placed

The author believes the mock rub aided him in collecting his trophy.

numerous drops of undiluted estrus doe urine. I then placed a few pieces of tree bark which I had allowed to soak in the urine for approximately a week. This would retain the scent much longer than the ground application.

Next I located a small tree at each end of the island. With the tree I made a scent pole and placed the strongest smelling estrus scent I had. I felt with the temperatures and the various other odors of the swamp, this was definitely needed.

Before exiting the swamp I prepared two mock scrapes. The mock scrapes were made quickly by clearing the debris from beneath a low hanging limb. With the debris removed, I sprayed the cleared area with undiluted buck urine.

Unlike some of the instructions I have read pertaining to mock scrapes, I prefer the use of buck urine. The reason for this is because it is more natural. Hunters should remember that it is the buck which makes the scrapes!

Now with my attractants in place, I headed for the next island which was approximately 50 yards beyond the one I had just prepared.

Upon reaching the second island I applied the same method of scent application as I had at the first island. This application had hardly been completed when I noticed the birth of another day. Now it was time for me to take my position for the hunt.

During my scouting of the area I had located a beaver house approximately halfway between the islands. This would serve as my stand site. I felt that with the aid of my binoculars I could spy on both islands.

I had just completed covering the boat with dried swamp grass when the first sounds of daylight began. All around me the sounds of ducks greeting the day filled the

Even in the refuge of the deepest swamps whitetail deer still go about their business of eating, sleeping, and breeding when the time comes. (Photo by Judd Cooney)

air. I listened to the quacking increase as I placed pre-cut cane in the beaver house.

The cane would serve to distort my human outline on the beaver house. Now with everything prepared I poured hot coffee from my thermos and relaxed.

I had only taken a few sips of coffee when I heard the sound of splashing water.

Within seconds my ears detected the exact location from where the sounds were being created. Confirming the location of the sounds, I realized something was wrong. The sounds were coming from directly behind me. This indicated the deer were not traveling as I had thought they would.

Quickly I located the deer with my binoculars as they bounced across the bed of swamp grass. With the increas-

Whitetail deer can be surprisingly aquatic—especially if pressured. Never overlook swamp islands, even if surrounded by considerable water. (Photo by Judd Cooney)

ing light of day, I could see that none of the deer supported antlers. My heart began to sink as I thought my plan was a sure failure.

Then suddenly, the sight of another the deer captured my attention. The antlers of the buck on the first island were clearly visible as the sun reflected from them. I wasted no time in viewing the buck with the crosshairs of my rifle scope. I also wasted no time in deciding I would shoot the buck if I could.

Minutes ticked away while I watched the buck scurry around the island. My excitement grew even more when while I watched the buck paw at the mock scrape I had prepared. The only thing that was allowing me to enjoy watching the buck fall prey to my trickery was two small trees shielding its shoulder. Now I could only wait for the buck to enter the small clearing in the middle of the island.

Time appeared to stand still as I waited and watched the deer through the rifle scope. Then I saw the buck suddenly raise its nose high in the air. With these movements I realized the buck had detected the scent from the scent pole at the other island.

I placed the scope on the small opening of the island. Now I could watch the buck react to the scent and wait for the moment of truth to unfold.

The few seconds it took the buck to travel to the opening seemed like an eternity. As I watched, my hands began to quiver with excitement. I knew that within seconds I would have to steady myself, and I took a deep breath.

Today as I write this, the memory of that buck shines in my thoughts. That massive 10-pointer is one of the top three bucks I have harvested in my lifetime. A dream which might not have happened if not for the use of scent.

Since the morning I collected my first swamp buck many more have fallen. There is no question some of the best trophies I have harvested have come from swamps. In hunting these areas, I have learned many things pertaining to hunting the magnificent whitetail deer. The greatest of all these lessons is to never underestimate what a deer can and will do.

I have seen deer swim across large bodies of water with ease. I have seen numerous deer avoid swimming small patches of water by walking over a beaver dam instead. The largest of these was a super-splendid 11-pointer. This buck's antlers are now part of my collection. This buck had been seen by hunters miles away feeding in a cornfield a week earlier.

So what is the moral to the story? Swamps are areas which receive little, if any, hunting pressure. During periods of high hunting pressure, swamps can become potential hot spots. This is true for hunting really big, mature bucks. These "super bucks" will be harvested by a hunter who is willing to go the extra mile and use the tools which are at our disposal.

In closing this book I would like to thank all my readers. In these final words I would also like to wish each and every hunter that their dreams may come true. I would also like to say that if any of these words achieve their purpose, then writing this book has been well worth the time I have spent.

May God Bless and Hunt *Safely*.

B.B.

INDEX

Note: Bold page numbers indicate illustrations or photos.

WHITETAIL SECRETS
VOLUME SEVEN — SCENTS FOR SUCCESS

Black and white photography by Bill Bynum

Color photography by Charles J. Alsheimer:
Pages 8, 22, 34, 50, 66, 76, 90, 102,
112, 122, 134, 146, 162, 178

Illustrated by Rick Carrell

Designed by Kirby J. Kiskadden

Text composed in Berkeley by
E. T. Lowe Publishing Co., Nashville, Tennessee

Color Separations and Film prepared by
D&T Bailey, Nashville, Tennessee

Printed and Bound by
Quebecor Printing, Kingsport, Tennessee

Text sheets are acid-free Warren Flo Book
by S. D. Warren Company

Endleaves are Rainbow Parchment by Ecological Fibers, Inc.

Cover material is Taratan II Bonded Leather by Cromwell